Contemplating Climate Change

Global climate change policy has failed us all, but what is the reasoning that underlies this failure? Why are some people more disposed to reflect on confounding issues like climate change, recognise the danger, seek a solution, and act accordingly, more than others?

This book is concerned with how we think and act in response to climate change. In particular, faced with deep uncertainty and the multifaceted complexities that characterise the climate change conundrum, how the various actors and institutions involved in the policymaking process make decisions that both aid and impede in the design and implementation of climate change policy. This book focusses on how these actors and institutions frame and use the knowledge available – under conditions of competing ideologies and interests – and synthesise it to form often-disparate mental models, or worldviews, that inspire them to become firm advocates of meaningful climate change action or indeed, sceptics that continue to downplay the threat, and hence the need for urgency. By exploring how we think about climate change and the disparate mental models we hold as a result, this book explores why humankind has thus far failed in its endeavours to solve the climate change problem.

This book will be of great interest to students and scholars of climate change, environmental policy and environmental psychology.

Stephen M. Dark is a Semester Teaching Fellow at Bond University, Australia, and holds a PhD in Environmental Epistemic Cognition.

Routledge Advances in Climate Change Research

Pricing Carbon in Australia
Contestation, The State and Market Failure
Rebecca Pearse

The Paris Framework for Climate Change Capacity Building
Mizan R. Khan, J. Timmons Roberts, Saleemul Huq and Victoria Hoffmeister

The Anthropology of Climate Change
An Integrated Critical Perspective
Hans A. Baer and Merrill Singer

EU Climate Diplomacy
Politics, Law and Negotiations
Edited by Stephen Minas and Vassilis Ntousas

The Global Climate Regime and Transitional Justice
Sonja Klinsky and Jasmina Brankovic

Climate Justice and the Economy
Social Mobilization, Knowledge and the Political
Edited by Stefan Gaarsmand Jacobsen

A Critical Approach to Climate Change Adaptation
Discourses, Policies and Practices
Edited by: Silja Klepp and Libertad Chavez-Rodriguez

Contemplating Climate Change
Mental Models and Human Reasoning
Stephen M. Dark

For more information about this series, please visit: www.routledge.com/Routledge-Advances-in-Climate-Change-Research/book-series/RACCR

Contemplating Climate Change

Mental Models and Human Reasoning

Stephen M. Dark

Routledge
Taylor & Francis Group
LONDON AND NEW YORK

earthscan
from Routledge

First published 2019
by Routledge

2 Park Square, Milton Park, Abingdon, Oxfordshire OX14 4RN
52 Vanderbilt Avenue, New York, NY 10017

Routledge is an imprint of the Taylor & Francis Group, an informa business

First issued in paperback 2020

British Library Cataloguing-in-Publication Data
A catalogue record for this book is available from the British Library

Library of Congress Cataloging-in-Publication Data
A catalog record has been requested for this book

ISBN: 978-1-138-60000-3 (hbk)
ISBN: 978-0-367-50769-5 (pbk)

Typeset in Times New Roman
by Wearset Ltd, Boldon, Tyne and Wear

For Srilai, Saraid, and Seabastian, without whom this book would not have been possible. And in memory of my mother, thank you for 'everything'.

Contents

Figures

1 Introduction

Contemplating climate change

> Thus *contemplation* enlarges not only the objects of our thoughts, but also the objects of our actions and our affections: it makes us citizens of the universe, not only of one walled city at war with all the rest. In this citizenship of the universe consists man's true freedom, and his liberation from the thraldom of narrow hopes and fears.
>
> (Bertrand Russell, *The Problems of Philosophy* ([1912] 2001, p. 93))

A preface to a crime

A disposition for contemplation, and the actions that flow from it, of the kind described by Bertrand Russell (1872–1970) above, is the essence of this story. More to the point, how certain thinking and acting, or lack thereof, can lead to the zombification of society and the greatest and most duplicitous crime of all – the demise of our planet. In this story, humankind is both the protagonist and the antagonist. Climate change is the object of interest – how we humans think and act in response to it. My intent was to establish why climate change policy has failed us all. The extensive cast of actors and institutions herein, that more, or less, contribute to this failure, was taken from the realms of science, politics and media, and economics and the publics. Explicitly, those actors and institutions engaged in the design and implementation of climate change policy. And actors they indeed are; our Mass Media Age would settle for nothing less.

It is presumed that these actors and institutions make climate change policy decisions and choices in response to empirical knowledge about the implications of climate change. The plot of this story revolves around how these actors frame and use this knowledge under conditions of competing ideologies and interests, synthesise it to form often-disparate mental models (or worldviews), and how these disparate mental models either aid or impede the design and implementation of climate change policy. Who prevails is the subject of monumental importance to planet Earth and all its inhabitants.

Climate change needs no introduction. Books, documentaries, and journal articles are replete with scientific data that bear testament to its scientific reality. The climate system is warming, and since the mid-twentieth century, human activity is the dominant cause (IPCC, 2013). Climate change inaction to this scientific reality is a contemporary and global problem of ominous implications. Despite the need for meaningful action, climate change mitigation and adaptation policy per se remains the subject of a spirited debate dating from the 1960s. The debate over human-induced climatic change, however, dates from the late 1800s. This is my story as to why global climate change inaction and its protracted inquisition persists in the face of such overwhelming validation.

This book is dedicated to students of learning – those with an inquisitive mind – from experts to non-experts. It is written from a Western perspective and is therefore limited by that perspective. It is purposefully interdisciplinary for the reason that climate change impacts on every aspect of life: succinctly, climate change is a complex environmental, cultural, social, economic, psychological, and political hazard of 'larger than life' proportions. A truly interdisciplinary approach is therefore required to successfully identify solutions and genuinely confront the climate change conundrum (Naznin, Hossain, & Huq, 2017). Notably, I have focussed on the thoughts of individuals more so than schools of thought, for the reason that schools of thought are in effect schools of groupthink in which like-minded people deliberate with one another in general agreement because they all think in much the same way (see for example Sunstein, 2000). Finally, if only one thing can be gleaned from this book, it is learning to think and reason for one's self – for diversity of thought is the true spice of life. Moreover, contrary to popular opinion, the future prosperity of humankind rests heavily on reinventing the wheel, probably several times over.

The investigation: a brief synopsis

Tim Mayers (2017), although not referring to climate change directly, sums up the climate change conundrum pretty much perfectly: 'Sometimes things are difficult to see when they are right in front of you. And sometimes, if the things themselves seem clear, the nature of the *connections between them* is difficult, if not nearly impossible, to see' (p. xi). Such is the nature of climate change.

Solving the climate change conundrum is indubitably complex and multifaceted (Incropera, 2015; Visschers, 2018; Young, 2017). Fathoming this complexity has inspired a mélange of competing worldviews. Competing worldviews about climate change have consequences for climate change policy. The successful design and implementation of global climate change policy are rare indeed. In light of this, I began by asking the question: what caused climate change policy to fail or what prevented it from succeeding? Early research pointed to two likely propositions: politico-media framing and the formation of disparate mental models – in effect, *nurture* (politico-media framing) versus *nature* (the mental models we form and hold). These two propositions arguably exist among multifarious others of considerable social, political, and economic bearing, but this book was not meant to be a comprehensive survey of all the variables that have derailed a meaningful global response to climate change. Nor was it meant to be a comparison of the multitude of theories that might have contributed to that derailment.

From the two likely propositions proffered foregoing, I developed a theoretical argument and deduced three hypotheses (or questions) to test my theory: (1) To what extent is the politico-media framing of climate change associated with the failure of climate change policy? (2) To what extent is the mental-modelling of climate change associated with the failure of climate change policy? (3) If there are relationships within and/or between (1) and (2) above, how do they combine to either aid or impede the successful design and implementation of climate change policy? I subsequently present an argument to support my theoretical conclusion to these three questions. I concluded that climate change policy per se fails, whether by design or implementation, for the same reason it succeeds: the human capacity and proclivity for a reflective mind mode which has a greater probability of validating an accurate mental model of the climate change problem and as such, acts as a blueprint for a commensurate action. I also submit the theoretical proposition that negativity bias is associated with a disposition for a System 2 mind mode, but this disposition is conditional upon, among other things, an individual's capacity and proclivity for reflective thinking and reasoning.

This investigation was conducted by way of theoreticians' theory. This was performed by arranging theoretical concepts as representative of social life. These concepts provide a vehicle for the generation of theoretical perspectives – a method to observe the social world from the perspective that best explains the proposed research theory. The theoretical perspectives comprise a number of ontological assumptions, such as the perceived dominance of rational choice over public choice. Theoreticians' theory is approached, broadly, from the worldview of C. Wright Mills (1959) and his conception of 'the sociological imagination'. C. Wright Mills was concerned with the limitations exacted on social research by the imposition of a dominant theory or method. He called this 'abstracted empiricism' and referred to abstracted empiricists as 'methodologically inhibited'.

In summary, in order to explain/resolve the maladies of the social world, Mills (1959) argued that imagination was needed to enable one to move from one perspective to another – 'from the political to the psychological; from examination of a single family to comparative assessment of the national budgets of the world; from the theological school to military establishment; from considerations of an oil industry to studies of contemporary poetry' (p. 7). I submit that solving the climate change conundrum, due to its multidisciplinary nature and inherent complexities, requires an ample dose of the sociological imagination.

To add structure to this story, a theoretical framework of sorts, a general model and a series of hierarchical theoretical models in the form of diagrammatic representations were constructed to explain the interrelationships between concepts and ontological assumptions. The models encompass the hypotheses about the objects in the model and are broadly constructed in the frame proposed by David Willer (1967). Willer's (1967) endeavour was to provide a 'methodological framework within which the imaginative sociologist could work' (p. vii).

The plot

Dark days approach. Global climate change policy has failed us all (see for example Doan, 2014; Dunlap & Brulle, 2015; Global Carbon Project, 2017; Head, 2014; Helm, 2010; Howlett, 2014; IPCC, 2014e; Jamieson, 2014; Pezzey, 2014; Roberts & Pelling, 2018; Rosenzweig, 2016; Stern, Bowen, & Whalley, 2014; Stoknes, 2015: Stoner & Melathopoulos, 2015). 'Policy failure' was defined by Michael Howlett (2014) as when a stated policy initiative does not correct or resolve a problem in purely programmatic terms, or when a policy innovation is not adopted in the first instance. Under this definition, in light of increasing CO_2 emissions and political procrastination, we can presume that global climate change policy has indubitably failed. As Ciplet, Roberts, and Khan (2015) observed:

> Despite decades of activity by nations shuttling around the world to give speeches and participate in negotiations, setting up entirely new multibillion-dollar carbon markets to 'efficiently' reduce emissions, and developing a series of new institutions and funding agencies to address this problem [of climate change], overall, things have continued to get worse.
>
> (p. 3)

In liberal democracies, public policy is the instrument of public opinion and politicians presumably work in the public interest. Thus, climate change policy is, theoretically, reflective of the general will of the people. Competing ideologies and disparate mental models of the climate change object, however, render this premise more complex than it might first appear. From the worldview of political pragmatism, climate change is a policy problem. In reality, however, it is a thinking and acting problem. Despite the need for a pragmatic approach, it cannot escape the complexity of its implications and the uncertainty of its prediction. As the former head of the U.S. Environmental Protection Agency (EPA), William D. Ruckelshaus (1992) remarked: 'The difficulty of converting scientific findings into political action is a function of the uncertainty of the science and the pain generated by the action' (p. 365). Nonetheless, scientific knowledge rests on its capacity to inform policy, and the efficacy of policy design and implementation is contingent on institutionalising the link between scientific knowledge and practice.

Policy is administered and enforced by the rule of law. Human activity operates within a legal and regulatory rule of law framework. This framework protects what is possible and endeavours to prevent what is not. Government bodies are charged with a 'duty of care' to ensure that the physical environment of communities and cities remains within acceptable parameters of safety and security, and that the built and natural environments enhance rather than detract from human wellbeing. With the advent of human-induced climate change, concern about the wellbeing of the Earth and its inhabitants has intensified. Environmental protocols have been established to allay this concern. The Precautionary

Principle, for instance, is widely viewed as a universal judicial standard for climate change prevention and accountability (Dark & Burgin, 2017). It is regularly cited in cases that involve development in eco-sensitive locations such as flood plains and vulnerable coastal zones.

According to Robin Attfield (2014), the Precautionary Principle first appeared in German and European legislation, then British, and later in other countries around the world. A version of the Precautionary Principle was in practice in Germany in the early 1980s (Boehmer-Christiansen, 1994), from which it entered into the English language as a translation of the German *Vorsorgeprinzip* (which pertains to the idea of enabling the government to take preventive action). The *Vorsorge* concept concerned 'the protection, even defence, of the environment or literally the world around us', and also advocated sustainable development (Boehmer-Christiansen, 1994, p. 31). Its approximate translation from German to English reads as follows:

> The principle of precaution commands that the damages done to the natural world (which surrounds us all) should be avoided in advance and in accordance with opportunity and possibility. Vorsorge further means the early detection of dangers to health and environment by comprehensive, synchronised (harmonised) research, in particular about cause and effect relationships ..., it also means acting when conclusively ascertained understanding by science is not yet available. Precaution means to develop, in all sectors of the economy, technological processes that significantly reduce burdens, especially those brought about by the introduction of harmful substances.
>
> (As cited in Boehmer-Christiansen, 1994, p. 37)

The meaning of 'precaution' in this instance is far-reaching. It not only involves acting in advance of a conclusively ascertained scientific understanding, but also developing (or modifying) all sectors of the economy (as a reference to sustainable development) and technological processes (to reduce harmful substances, such as CO_2 emissions).

It is presumed that Germany's *Vorsorge* concept was the prototype for the United Nation's more contemporary version, the Precautionary Principle. The Precautionary Principle emerged from the United Nations Conference on Environment and Development held in Rio de Janeiro in 1992. This was a gathering of minds with the stated intent, among others, to establish a protocol for environmental protection. It is worded as follows: 'Where there are threats of serious or irreversible damage, lack of full scientific certainty shall not be used as a reason for postponing cost-effective measures to prevent environmental degradation' (Rio Declaration on Environment and Development, 1992, Principle 15). It was operationalised under the premise that:

> The guidance which science provides will seldom be so precise as to remove all uncertainty. In matters affecting our survival, we cannot afford to wait

> for the certainty which only a post-mortem could provide. We must act on the Precautionary Principle guided by the best evidence available.
>
> (Strong, 1992, n. p.)

The Precautionary Principle has been adopted in many countries, although there are variations in its wording. There is, however, no consensus regarding a universal interpretation of the Precautionary Principle, nor a consensus that one is possible (Steel, 2014). For instance, the Precautionary Principle provides no clear guidance as to what degree of proof is necessary before its application is triggered. Gerry Bates (2013), a specialist in environmental law and policy, has suggested a conciliatory measure of probability (<0.05) as a way of legitimising the Precautionary Principle to a scientific standard. Scientific standards of probability, however, have little to do with social justice, moral accountability, and political feasibility (see for example Patterson et al., 2018).

Obfuscation surrounding the application of the Precautionary Principle is not new. In the mid-1970s Adrian Deville and Ronnie Harding (1997) argued that the Precautionary Principle emerged with no clear rules as to its use; to date, both the public and private sectors continue to grapple with a concrete framework for its application. Andrew Jordan and Timothy O'Riordan (1999) highlighted its inadequacies. They asserted that the Precautionary Principle was not a 'well-defined principle nor a stable concept but the repository for a jumble of adventurous beliefs that questioned the status quo of political power, ideology, and environmental rights'. They further maintained that the concept had only marginal coherence except that it challenged the 'authority of science, the hegemony of cost–benefit analysis, the powerlessness of victims of environmental abuse, and the unimplemented ethics of intrinsic natural rights and intergenerational equity'.

The 'jumble of adventurous beliefs' referred to by Jordan and O'Riordan (1999) above, might allow 'the rule of thumb' to be applied to the decision-making process rather than the rule of law based on empirical evidence. In this way, the Precautionary Principle might be interpreted as a heuristic for decision-making when some of the outcomes of the decision are unknown. Heuristics, commonly defined by their departure from scientific thinking (Osbeck, Nersessian, Malone, & Newstetter, 2010), sometimes yield reasonable judgements but might also lead to critical and systematic errors (Crupi & Elia, 2017; Kahneman & Tversky, 1973; Norman et al., 2017). This is because human judgement depends on one's conceptual grasp of things in order to competently reason about the relations between them. In short, cognitive biases (heuristics) should not be confused with empirical evidence, which presents a tricky proposition.

In order to reform the Precautionary Principle, environmental health, which is analogous to human health, should be prioritised (Dark & Burgin, 2017). That is, the Precautionary Principle cannot achieve legitimacy if it is merely one option in a list of possible options arbitrarily subject to the discretionary powers of the politico-elite. Preserving the environment, and the associated ecosystem services that it provides, should be elevated to a position of precedence against which all other competing goals, such as economic growth, are measured in terms of their negative/positive impacts on the short- and long-run welfare of planet Earth and its inhabitants. Indeed, the efficacy of the Precautionary Principle is diminished by the uncertainty of the scientific framework within which it is contrived to operate, and the continued disputation over the short- and long-run socio-economic costs and benefits associated with its design and implementation. Averting dangerous climate change should not be subject to the whims of discretionary powers.

Humans and the weather: the early years of discovery

Despite the human endeavour to control nature, to mould it to a preferred state of affairs, there is no escaping humankind's often-stormy relationship with the weather. Edwin Grant Dexter (1899), for instance, reported that on cloudy and rainy days there were fewer bad marks given in schools, fewer arrests for assault, and even fewer suicides than on clear, dry days. Recent research by Zivin, Hsiang, and Neidell (2018) revealed that short-run changes in the weather resulted in a statistically significant decrease in math performance, but had no effect on reading. In contrast, long-run changes, in reference to climate change, appeared to have no effect on cognitive capacity. This suggested that compensatory behaviour (adaptation) was occurring.

The effect of the weather on mood has been extensively explored, revealing different results. Sufferers of Seasonal Affective Disorder (SAD)[1] experience recurrent episodes of major depression to mania and hypomania in parallel with the changing seasons (American Psychiatric Association, 2013). Major depression is regularly experienced in the winter, for instance, then diminishes or disappears with the arrival of spring. Although the weather's effect on mood pales in significance against the effect of climate change on planet Earth, it does highlight the complex and chaotic nature of our relationship with the weather, and what a dramatic change in seasonal weather patterns might exact on the human condition.

Climate curiosity can be traced back to classic Greece. In the third century BC, Theophrastus, a pupil of Aristotle's, noted that the permanent draining of the marshes around Larissa in Thessaly resulted in greater extremes of cold and an increase in freezings (Glacken, 1967). In the Middle Ages, the effects of deforestation on climate were discussed by Albert the Great. Discussions of this kind spread to North America (post-European settlement) where travellers believed they observed a warming of the climate in the aftermath of newly discovered lands that had been subjected to deforestation. Research by Kelleyanne Malinen and Chérif Matta (2018) has shown that an early reference to climate change appeared in a book entitled *The Emigrant* (1846), in which Sir Francis Bond Head (1793–1875) noted that 'forest wildfires were changing the climate of North America'.

It is likely that the Austrian geographer, Eduard Brückner, became aware of the possibility of climate change in the mid-1880s (Stehr & von Storch, 2000).

In 1888 Brückner presented his initial findings, entitled *Is Our Climate Changing?*, at a public gathering at the University of Dorpat (now the University of Tartu in Estonia). He speculated on how climate variability would have serious socioeconomic consequences. Migration patterns, the harvesting of crops, the balance of trade between countries, and the political predominance of some countries would be disrupted. The following year Brückner presented these findings to a congress of geographers in Berlin. His work on climate change culminated in 1890 with the publication of an expansive discussion on climate fluctuations in historical times.

Central to Brückner's conjecture was, if forests increase precipitation then deforestation must necessarily decrease precipitation (von Storch & Stehr, 2006). He thus established the link between forests and anthropogenic climate change. Brückner's persistence that deforestation would cause changes in climate, which in turn would have implications for rivers, lakes, and agriculture, made him a central figure in early debates about global climate variability. The debate has endured to this day as deforestation persists. It has become a critical issue in the relationship between CO_2 emissions and predicted climate change, and the preservation of ecosystems as a means to averting such an event. Brückner's was an important voice in the recognition of climate change and its future implications, and he was perhaps the first protagonist with respect to endeavours to institute a deliberative climate change policy.

A central actor in the investigation of the properties of elementary and compound gases was Irish scientist John Tyndall. Tyndall experimented with the absorptive properties of gases and vapours. He theorised that different gases found in the Earth's atmosphere absorbed different amounts of radiant heat. This theory was presented more than 30 years earlier by French physicist and mathematician Joseph Fourier (1768–1830) and recorded in his important work *Théorie Analytique de la Chaleur* (1822). Tyndall (1861) concluded that 'Great differences have been shown to exist among gases as to their powers of absorption, and precisely similar differences as regards their powers of radiation' (p. 33).

Based on his experiments, Tyndall posited that elementary gases, such as 'hydrogen, oxygen, and nitrogen' absorbed less energy (radiant heat) than molecules of gases such as methane, carbon dioxide, and water vapour; with the latter offering scant resistance to the passage of radiant heat. At a meeting of London's Royal Society in 1859, Tyndall delivered what was considered to be the first public presentation of what is now known as the 'greenhouse effect' (Hulme, 2009). This research became pivotal to a subsequent understanding of the atmosphere's energy budget.[2] It is important to add here, that in 1856, Joseph Henry presented a paper to the American Association for the Advancement of Science written by American scientist Eunice Foote, which predated Tyndall's presentation. Her paper, entitled 'Circumstances Affecting the Heat in the Sun's Rays' (1856), made reference to the effect of air density on the thermal action of the sun's rays. Although Tyndall's experiments appear more comprehensive, it is likely that Eunice Foote was the first person to speculate on the probability of a greenhouse effect. The fact that she did not present her findings herself was symptomatic of the gender-based oppression of the times.

The accredited discoverer of how sensitive the Earth's climate was to changes in atmospheric composition was Swedish physicist and physical chemist Svante August Arrhenius (1859–1927). Drawing on the work of James Croll and John Tyndall, Arrhenius calculated that the global mean surface temperature would fall by approximately 5°C if CO_2 concentrations in the atmosphere were halved. Conversely, he calculated that the global mean surface temperature would rise by approximately 5–6°C if they were doubled. This is known as equilibrium climate sensitivity (ECS), and remains one of the major unknowns in climate science (Cox, Huntingford, & Williamson, 2018). That is, an instantaneous doubling of CO_2 concentrations in the atmosphere would equate to what rise in Earth's mean global surface temperature, over what period of time?

Although Arrhenius's presentation of his findings to the Stockholm Physical Society in 1895 was widely accepted, the climate change debate was only just beginning. Building on the work of Arrhenius, and adding fuel to the debate, was Guy Stewart Callendar, a British mechanical engineer. Callendar presented a paper to the Royal Meteorological Society in 1938 linking the concepts of the physical theory of CO_2 and the greenhouse effect, the rising concentration of CO_2 in the atmosphere, and the increase in global temperature to anthropogenic climate change. His paper was received with some scepticism of which he anticipated:

> Few of those familiar with the natural heat exchanges of the atmosphere, which go into the making of our climates and weather, would be prepared to admit that the activities of man could have any influence upon the phenomena of so vast a scale.
>
> (Callendar, 1938, p. 223)

Callender's warning was ominous. He also expressed a hint of disbelief himself at what he had discovered by pointing to the likelihood that one might indeed be looked upon as 'irrational' if he or she should admit that human activity could actually alter something so immense as the Earth's climate system. Suffice to say, he was right to be concerned.

According to the IPCC (2013) there has been a 40 per cent increase in CO_2 levels since pre-industrial times (1850–1900). Global CO_2 emissions have increased every decade from the 1960s to 2007–2016 (Global Carbon Project, 2017). The main source of emissions is the combustion of fossil fuels, and to a lesser extent from net land use change emissions. About 30 per cent of the emitted anthropogenic CO_2 has been absorbed by the oceans, causing acidification. In the early 1960s, climate modelling indicated that doubling the CO_2 in the atmosphere could result in a temperature rise of 10°C or higher (Möller, 1963). The last time the Earth experienced a period when global mean surface temperatures were around 10°C above the present level was during the Eocene epoch ≈55–34 million years ago. The Earth was free of ice and alligators roamed among the palm trees near the North Pole (Weitzman, 2012).

The historical journey of climate change probably came of age in a speech by British Prime Minister Margaret Thatcher to London's Royal Society in 1988. Thatcher (1988), while declaring her commitment to science and to the environment, expressed a deep concern that recent findings had raised the fear 'that we are creating a global heat trap which could lead to climatic instability' and the possibility that 'we have unwittingly begun a massive experiment with the system of this planet itself' (n.p.). Thatcher was a trained scientist – having studied chemistry at Oxford University – and was thus able to reconcile the uncertainty in the science and put aside the politics to accurately synthesise a mental model of what was transpiring and what might come.

The world according to the IPCC

Perhaps coincidentally, in the same year of Margaret Thatcher's speech to the Royal Society in 1988, the Intergovernmental Panel on Climate Change (IPCC) was established by the United Nations Environment Programme (UNEP) and the World Meteorological Organization (WMO). Its purpose was to provide the world with impartial scientific knowledge pertaining to the current state of climate change and its potential environmental and socioeconomic implications. The Anthropocene of climate change had officially arrived.

The IPCC advocates an 'inclusive, democratic, and objective' process of environmental assessment. In effect, it proffers a universal guide to climate change policymaking and assumes the role of the institutional interface between science and politics. It is apt to point out that the IPCC does not conduct any research of its own (actual experimentation or climate modelling per se) nor does it actively monitor climate change parameters, but aims to comprehensively assess the latest scientific literature worldwide and provide recommendations and options for policymakers.

The mainstay of the IPCC's predictions is a probability range (see for example IPCC, 2013, p. 36), which is used to indicate the likelihood of climate change events. The use of such wide-ranging probabilities, however, might add to the uncertainty. For instance, it is not clear how to synthesise the IPCC's probability scale into an intellectually valid mental model that represents the prospect of surviving an extreme climate event, and there is no obvious statistical method for the publics to test these probabilities.

A study of how the public interprets the probabilistic statements (in the IPCC's 2007 AR4 report) by Budescu, Broomell, and Por (2009), found that there were substantial differences in the way people understand the IPCC's set of probability terms in the context of their global interpretational guidelines. Even when respondents had gained access to these guidelines their judgements deviated substantially from them. They concluded that the IPCC's method was likely to convey levels of imprecision that were too large. The result of this study showed that the range of translations of probability (in terms of a percentage) to their verbal equivalents (for example, *very likely*, *likely*, *unlikely*, and *very unlikely*) varied significantly in the way individuals interpreted and used these probability expressions. In short, it is likely that the conversion of a probability to its equivalent in layman's terms is often lost in translation. The probabilities used by the IPCC encompass wide-ranging possible outcomes that might be progressively changed in accord with updated climate predictions.

Probability, as a calculation of an event occurring, should not be confused with belief. People still ask, 'do you believe in climate change'? Politicians, for example, regularly frame climate change policy in the context of 'we believe that climate change is happening'. Immediately the element of belief is brought to bear; such as do you believe in God? But global warming is factual – both past and present. So the question should have progressed from 'is climate change happening and is it harmful', or 'what is the probability of climate change being anthropogenic', to what is the 'survival probability index of humankind under the current fossil-fuelled regime', or 'what is the survival probability index of humankind under a clean energy regime'?

Surviving the long-term implications of climate change are not clear. The foremost international body for the assessment of climate change, the IPCC, states that: 'Warming of the climate system is unequivocal, and since the 1950s, many of the observed changes are unprecedented over decades to millennia' (IPCC, 2013, p. 4). 'It is *extremely likely* that human influence has been the dominant cause of the observed warming since the mid-20th century' (IPCC, 2013, p. 17). 'A changing climate leads to changes in the frequency, intensity, spatial extent, duration, and timing of extreme weather and climate events, and can result in unprecedented extreme weather and climate events' (IPCC, 2012, p. 7). 'Due to sea-level rise projected throughout the 21st century and beyond, coastal systems and low-lying areas will increasingly experience adverse impacts such as submergence, coastal flooding, and coastal erosion (*very high confidence*)' (IPCC, 2014e, p. 16). These statements surely must dramatically alter our current worldview and its everyday realities. A new climate system is increasingly imposing itself – one that is much less human-friendly than the current one, and will become more so. But what exactly does this mean?

Waking up to a complex climate change

Climate change has become a metaphor for complexity. Since about 1970 the Earth has been in a radiative imbalance. 'It is *virtually certain* that the Earth has gained substantial energy from 1971 to 2010' (IPCC, 2013, p. 39). This imbalance in the Earth's albedo[3] has led to what is known as the greenhouse effect. Greenhouse gases, primarily water vapour (H_2O), carbon dioxide (CO_2), nitrous oxide (N_2O), methane (CH_4), and ozone (O_3), trap heat within the surface-troposphere system causing global mean surface air temperatures to increase, referred to as global warming. This energy imbalance has implications for all life on Earth. As posited by the scientist and futurist, James Lovelock (2007b), in his Gaia hypothesis: the Earth is a single self-regulating organism and as such, 'The planet we live on has merely to shrug to take some fraction of a million people to their death' (p. 1). An *aide-mémoire* of the havoc that would transpire if we were to flagrantly ignore the signs of an ailing planet entombed in an interconnected cycle of decay.

It is evident that policymaking institutions have failed to keep pace with the complexities of contemporary interconnected issues, such as the multifaceted ways in which climate change impacts on society, and are in need of reform (Ciplet et al., 2015). Traditional models are subject to 'path dependency' – the future options available to them being a function of past options taken, and a legacy of past decisions that presuppose stationarity – the mistaken convention in time series methods that data are stationary. In a critique of static models in the context of climate change adaptation policy formation, the Commonwealth Scientific and Industrial Research Organisation (CSIRO; 2012) observed the following:

> Adaptation to climate change is a complex systems topic that, like sustainable development, is not adequately analysed in a conventional economic framework that is implicitly constrained by equilibrium thinking. Issues such as non-stationarity in the operating environment, thresholds and nonlinear change, and emergent properties across scales cannot be addressed from within this framing.
>
> (p. 3)

A defining feature of complex systems is their nonlinear behaviour (Young, 2017). They exhibit changes that are difficult to anticipate, sometimes transformative, and often abrupt. In contrast, mainstream thinking is dominated by models of gradual and incremental processes about environmental problems like climate change. There is the propensity to believe that, in the normative sense of the everyday, what happened yesterday will happen tomorrow. This makes past experience superfluous as a guide to developing future plans for effective environmental governance. The non-stationarity and nonlinear nature of climate change, then, has repercussions for those individuals possessed of a static mental model that is out of date. Jürgen Scheffran (2015) has forewarned that as a consequence of nonlinear effects, an increase above a certain global temperature might trigger instabilities that overwhelm the viability of both natural and social systems. This has consequences for seaside communities holding to 'lifestyle mental models' deficient in the implications of extreme climate events and the potential for a relatively rapid change in the climate.

According to the IPCC (2012), 'The nonlinearity of the climate system might lead to abrupt climate change, sometimes called rapid climate change, abrupt events, or even surprises' (p. 556). More precisely, geologically rapid changes in global temperatures are referred to as Dansgaard/Oeschger events. For example, about 70,000 years ago global temperatures rapidly declined plunging the Earth into the last glacial period. This comparatively cold period began to recede around 14,000 years ago. About 10,000 years ago, over an exceptionally short period of only 50 years, global temperatures increased by as much as 7°C, signalling a precipitous end to the last ice age.

Weather and climate are archetypal examples of complex systems. Simply stated, complexity is 'a condition of nonlinear and/or recursive relationships between causes and effects' (Earnest, 2015, p. 32). Moreover, 'Complexity limits the ability of individuals to identify the full set of possible outcomes or assign the probabilities to particular outcomes of specific actions' (Poteete, Janssen, & Ostrom, 2010, p. 58). For instance, natural resources, ecological systems, and socioeconomic and political systems all have attributes that affect the capacity of resource users to recognise how their actions might impact the resource's condition. These complex interactions have potentially destabilising effects which could lead to social disruption, urban violence, or even war.

The complexity of the interaction and interconnectedness between natural and social systems is often explained by chaos theory. Chaos theory largely owes its discovery to atmospheric science. The 'butterfly effect' is a term used to describe how seemingly minuscule perturbations in the atmosphere might ultimately give rise to significantly larger or even catastrophic events. It derives from a paper by Edward Lorenz (1972) entitled *Predictability: Does the Flap of a Butterfly's Wings in Brazil Set off a Tornado in Texas?* To emphasise the weather's unpredictability, Lorenz posed the hypothetical scenario: can the flap of a butterfly's wings in Brazil cause a tornado in Texas? Can two divergent events evolve into two events that differ by as much as the manifestation of a tornado?

That is, is the behaviour of the atmosphere unpredictable to the extent that perturbations of such small amplitude can result in such extraordinarily contradictory outcomes? Lorenz argued that miscalculations in weather prediction were inevitable and came from an incomplete coverage of weather systems, an incomplete knowledge of the underlying physics, and the fact that approximations were unavoidable. He pointed out that 'The atmosphere is not a controlled experiment; if we disturb it and then observe what happens, we shall never know what would have happened if we did not disturb it' (Lorenz, 1972, p. 1). Human-made climate change falls into this category.

Climate change is both a natural and human-made object of a particularly frustrating nature. It routinely attracts the adjectives 'catastrophic', 'cataclysmic', and 'apocalyptic'. Described by Deborah Curran (2009) as a 'classic wicked problem'; Levin, Cashore, Bernstein, and Auld (2012) as a 'super wicked problem'; and Ross Garnaut (2008b) as a 'diabolical policy challenge'. These dramatic expressions infer tipping points that could potentially trigger an abrupt transformation of the climate. Tipping points in this instance are predetermined thresholds at which the damage caused by climate change is irreversible and a feedback loop occurs which precipitates further warming.

Large and abrupt climatic changes have occurred in the past. Conceivably on a scale depicted in the 2004 movie *The Day After Tomorrow*[4] – although in terms of time scale, weather change was substituted for climate change. Nonetheless, as oceanographer and climatologist Stefan Rahmstorf (2001) explained, 'High-resolution paleoclimatic records from ice and sediment cores and other sources have revealed a number of dramatic climatic changes that occurred over surprisingly short times – a few decades or in some cases a few years' (p. 1). Indeed, climate change exposes the Earth to tipping points that might well be reached before the reality of the new world we are creating is acknowledged; thus, pushing it beyond the resolve of human ingenuity to remedy.

The IPCC (2013, p. 27) has estimated that to prevent the global mean surface temperature exceeding the internationally recognised tipping point of 2°C (relative to the period 1861–1880) with a probability of >66 per cent would require cumulative CO_2 emissions from all anthropogenic sources to remain between '0 and about 1000 GtC (3670 $GtCO_2$)' (1 Gigatonne of carbon = 1 GtC = 1015 grams of carbon, which corresponds to 3.667 $GtCO_2$; p. 27). Estimates of this type vary according to the climate model and assumptions used, how methodological issues are managed and resolved, and as new data come to light. Recent research has indicated that post-2050 most tropical regions will experience a climate departure from their historical range of variability, meaning that the mean climate will constantly exceed the historical bounds of variability (Mora et al., 2013). Other research indicates a comparatively shorter timeframe. Ecologist Lesley Hughes (2014), for instance, has used a 'business as usual' model to describe a scenario in which the world was on track to use up its carbon budget by 2028 and thus theoretically, breach the 2°C tipping point as a consequence.

An important point to note here is that the 2°C threshold is often attributed to climate economist William D. Nordhaus who purportedly proposed it in the mid-1970s. There is, however, no evidence to support this claim and no evidence to indicate that 2°C was an empirically defined tipping point. Despite this, it gained acceptance from the mid-1990s and was endorsed at the COP15 (15th United Nations Climate Change Conference of the Parties) held in Copenhagen in 2009. Contrary to its current official use, several leading climatologists have repeatedly warned that setting the threshold at what is essentially an arbitrary value of 2°C was irresponsible: things could go awry well before reaching this limit. Although 2°C has served as a useful reference point, as Nordhaus noted, it suggests stationarity whereas the climate system is exceptionally dynamic.

The 'global carbon budget' is the estimated amount of CO_2 emissions that can be emitted while still maintaining the likely probability of limiting the mean global surface temperature rise to 2°C above pre-industrial levels. The world has already warmed very near 1°C since pre-industrial times and the inertia in the climate system will potentially add a further 0.5°C (Hughes, 2014). It is predicted that breaching the supposed inflection point of 2°C would cause an irreversible climate shift. What is more, even if the atmospheric composition was held fixed at today's values or even if greenhouse gases were reduced, the IPCC (2013) has projected that ocean warming and sea level rise would continue for centuries.

Sea level rise is an ominous manifestation of climate change (see for example Chang et al., 2018; Goodell, 2017; Hardy & Hauer, 2018). Sea level is measured relative to land surface and hence, any variation reflects either a change in the position of the sea surface or a change in the position of the land surface, or both. Global mean sea level (GMSL) is currently rising at close to 3.0 mm per year (Dangendorf et al., 2017; Dieng, Cazenave, & Ablain, 2017). Sea levels have fluctuated throughout geological time, periodically advancing and withdrawing across coastal plains and continually reshaping the coastline (Blainey, 2015). In more recent geological time, some 18,000 years ago, a seminal event occurred. Massive swathes of ice near the north and south poles began to melt as the Earth warmed. The sea moved inland and most coastal dwellers lost their homelands. The Earth continued to progressively warm and sea levels continued to rise until about 8000 years ago. By this time, dinosaurs had been extinct for about 65 million years, Australia's Great Barrier Reef had formed, and temperatures were probably much the same as they are today.

Predicting future sea level rise is a hazardous affair. Currently, the largest contribution to sea level rise has been from ocean thermal expansion (The Royal Society, 2017). However, major uncertainties relate to the Greenland and Antarctic ice sheets (Gwyther et al., 2017). These massive ice formations have the potential to raise the global mean sea level up to 70 metres above current levels if completely melted (Alley, Clark, Huybrechts, & Joughin, 2005). The great ice sheets of Greenland hold about 10 per cent of the Earth's ice mass. The East and West Antarctic ice sheets in particular, comprise about 90 per cent of the Earth's ice and about 70 per cent of its freshwater (Gwyther et al., 2017). Allison, Alley, Fricker, Thomas, and Warner (2009) estimated that a complete melting of the ice sheets would equate to a potential global mean sea level rise of about 65 metres, comparable to Alley et al. (2005). Similarly, Gwyther et al. (2017) placed this figure at about 60 metres above current levels. However, under moderate warming scenarios, London's Royal Society (2017), in concurrence with the IPCC's (2014e) AR5, has estimated that ice sheet instability will not contribute more than several tenths of a metre by 2100.

The principal unknown in global mean sea level predictions is the potential for the rapid dynamic collapse of these ice sheets. This is compounded by the nonlinear correlation between the mass balance of the Greenland and Antarctic ice sheets, and the warming of the Earth (Mikkelsen, Grinsted, & Ditlevsen, 2018). Although data from the GRACE satellites mission (Gravity Recovery and Climate Experiment) launched in 2002, have provided greater insight into how water and ice are distributed around the Earth, and advanced computer modelling and measuring techniques, in particular, glacial isostatic adjustment (GIA), are continually improving, ice sheet mass balance[5] remains unclear (Davis & Vinogradova, 2017; Hansen et al., 2016). Geologic records indicate that due to past natural climate factors, global mean sea level has been highly variable. Relative to present levels, it ranged from as much as 5.5–6 metres higher during the last interglacial period (≈125,000 years ago; Dutton & Lambeck, 2012), to a peak of about 130 metres lower than present levels during the last glacial period (constrained to 26,500–19,000 years ago, as the most recent period when ice sheet volumes were at their combined maximum; Clark et al., 2009).

According to Kopp, Simons, Mitrovica, Maloof, and Oppenheimer (2013), polar temperatures were about 3–5°C warmer during the last interglacial warm period than they are today, and there is a 95 per cent probability that sea levels peaked at least 6.6 metres higher. Mean sea levels also varied considerably between geographically diverse sites. Perhaps coincidently, Quiquet, Ritz,

Punge, and Melia (2013) predicted a 3–5°C warming in Greenland by the end of this century. At the high end, the IPCC (2013) predicted a *likely* (66–100 per cent) global mean surface temperature increase of 2.6°C to 4.8°C for 2081–2100 relative to the reference period 1986–2005. This suggests a 66–100 per cent probability of a global mean surface temperature increase of 3.7°C by the end of this century. Suffice to say, this dramatically exceeds the internationally recognised tipping point of 2°C above pre-industrial levels.

In consideration of this, research by Hansen et al. (2016) has indicated that ice melt and extreme storms in the prior interglacial period, in which temperatures were less than 1°C warmer than they are today, correlated with a global mean sea level rise of 5–9 metres. Dutton et al. (2015) concurred, noting that the present warming of the Earth is approaching a level that coincides with substantial polar ice melt that occurred in the past. Small increases in the mean global temperature during recent interglacial periods, relative to temperatures in the pre-industrial period, translate to a global mean sea level rise of ≥6 metres. In contrast, the IPCC (2014b) has predicted that global mean sea level was unlikely to rise more than one metre in the current century. Of significance, however, is that current anthropogenic climate forcing is 'stronger and more rapid' than the paleoclimate forcings of prior interglacial periods (Hansen et al., 2016).

Although natural variability remains the dominant climate mode, climate change pushes both weather and climate to extremes. It is not a question of whether an extreme climate event is caused by climate change. All weather events are more or less influenced by climate change because the natural and human-made environment in which they occur is now warmer and moister than it was in the past (Trenberth, 2012). This 'new climate system' can be largely attributed to the concentration of greenhouse gases in Earth's atmosphere which has reached a level that is outside the natural variability experienced for at least the last million years (Steffen, 2014). The natural and anthropogenic components that cause shifts in the climate can be measured, but what fuels debate and fires contention is that they cannot be 'precisely' measured independent of one another. This lack of precision gives rise to uncertainty.

Decision-making under conditions of deep uncertainty and denial

> Ambivalence, ambiguity, equivocality.... These words convey the feeling of mystery and enigma; they also signal trouble, whose name is uncertainty, a dismal state of mind, called indecision or hesitation.... We are inclined to believe that we would feel much safer and more comfortable if situations were unambiguous – if it were clear what to do and certain what would happen if we do it.
>
> (Zygmunt Bauman, *The Individualized Society* (2001, p. 109))

Polish sociologist and philosopher Zygmunt Bauman (1925–2017), defines a well-established point: humans are decidedly uncomfortable with decision-making under conditions of uncertainty. It messes with their endeavours to predict the future, as Bauman duly noted above. Although Laplace extended on the Newtonian paradigm of an orderly universe through probability theory, Heisenberg's Uncertainty Principle arguably dispelled any declarations of precise prediction. This was based on the premise that although there are general boundaries to most phenomena, within these boundaries exact outcomes are uncertain. Even so, the pursuit of order and the inherent desire to comprehend and predict our world and its condition has been a major preoccupation throughout recorded history. From the Delphi Oracle to supercomputers, the endeavour to foresee and foretell has captured the imagination of writers and movie-makers alike. So, although uncertainty succeeds in confounding our interactions with the world, it is nonetheless accepted as a universal truth of natural and human phenomena.

Uncertainty is the anomaly that the future does not exist until it is the present. Scientific uncertainty is fundamental to climate change predictions and a consequence of natural variability, model and scenario uncertainty, and an incomplete knowledge of Earth system dynamics (Latif, 2011). More precisely, scientific uncertainty stems from three types of manifestations: statistical variability and heterogeneity (or aleatory uncertainty); model and parameter uncertainty (or epistemic uncertainty); and deep uncertainty (or uncertainty about the fundamental processes or assumptions underlying a risk assessment; Institute of Medicine, 2013).[6] For example, in a perfect world, scientists, politicians, and the publics, would have perfect information – precise, complete, and uncontested – that would allow for accurate analyses and decision-making leading to measurable benefits. Decision-makers, however, have imperfect information, and cannot accurately predict the consequences of their decisions, which causes varying degrees of discomfort – from mild stress to acute anxiety.

The term aleatory derives from the Latin *alea*, which means 'the rolling of dice'. An aleatoric uncertainty is thus one characterised by the inherent randomness (based on a calculated probability) of a phenomenon occurring (Der Kiureghian & Ditlevsen, 2009). Aleatory uncertainty (variability and heterogeneity, or randomness) is often represented in environmental conditions, levels of exposure, and an individual's susceptibility. Statistic-based probabilistic risk analysis is used to quantify aleatory uncertainties.

Natural variability plays a role in climate change prediction. Natural variability in the climate system, as opposed to human-induced climate change, is often a point of contention. Subsequently, the IPCC adopts a probability scale to quantify the likelihood of a climatic event: '*Virtually certain* 99–100% probability, *Very likely* 90–100%, *Likely* 66–100%' etc. (IPCC, 2013, p. 955). As a consequence of always-imperfect information, predictions of future events can never be 100 per cent accurate except by chance. This allows climate change denialists, in particular, to claim that the science is not 100 per cent certain. Although this is true, indecision should never be allowed to reign supreme, as this would result in chaos. Mindful that chaos is also the mother of invention, rather than necessity. 'Necessity', as it were, lacks resolve and inspiration.

The term epistemic is derived from the Greek *episteme* which means knowledge. Hence, an epistemic uncertainty is caused by a lack of knowledge (or data; Der Kiureghian & Ditlevsen, 2009). More explicitly, epistemic uncertainty concerns model and parameter uncertainty, which requires decision-making in the absence of perfect knowledge. It entails decisions about the most appropriate statistical model and what degree of expert elicitation should be used in the specific analysis. It also entails an assessment of technological parameters such as what technologies are appropriate and available. Most contentious is that it requires predictions about what technologies might be available in the future.

Expert elicitation is invariably an input to predictions made about climate change by the Intergovernmental Panel on Climate Change (IPCC). Whether the parameters of the modelling underpinning those predictions should include technologies such as carbon sequestration, or geoengineering, based on what is known, are questions to be answered and decisions to be made. Notably, there is a commonplace argument that yes, climate change is happening, but human ingenuity through science and technology innovation will find a fix. However, there is a lack of empirical evidence to qualify this argument. This brings to the fore the prospect of the 'technological singularity'.

Mathematician and science fiction author, Vernor Vinge (1993), argued that within 30 years humankind would have the capacity to create 'superhuman intelligence' for the betterment of the human-made world – or perhaps for its decline. In a similar but more positive vein, futurist Ray Kurzweil (2005) predicted that the exponential progression of technology would ultimately provide a universal panacea to the world's foremost maladies. Suffice to say, deep uncertainty surrounds predictions of this kind. The Slovenian philosopher, Slavoj Žižek (2009), for instance, citing Wilfredo Pareto's 80–20 principle, declared that advances in technology as a means to boosting productivity has the potential to render 80 per cent of the world's labour force 'irrelevant and of no use' – only 20 per cent of the labour force would be required to do the necessary work. Subsequently, any political system that would endorse the prospect of rendering 80 per cent of the labour force redundant is 'itself irrelevant and of no use'. It is apt to pause here to consider Žižek's warning. The irony of stridently pushing ahead with technology innovation while deferring efficacious action to avert the risk of dangerous climate change. Unless of course we invoke the technological singularity as the impending means to an end.

In the scientific reality of the present, deep uncertainty remains an unremarkable but foreboding challenge (Ansell & Boin, 2017; Beck, 2015; Nair & Howlett, 2017). Traditional analytic approaches that focus on the probability of certain outcomes, based on a set of parameters and assumptions, no longer apply because there is minimal consensus as to the parameters and assumptions of the model on the part of the decision-makers involved. Climate change is conceivably the most esoteric environmental problem to confront humankind. Deep uncertainty exists about the future costs and benefits of actions to effectively address the enormity of the threat. It is evident that technology innovation is required to play a role, but under conditions of ill-defined risk and uncertainty, what decisions and actions should policymakers make and take to inspire such innovation?

Scientist and scholar Robert J. R. Lempert (2002, 2015; Lempert & Groves, 2010; Lempert, Groves, Popper, & Bankes, 2006), a prolific author and researcher of decision-making under conditions of deep uncertainty, has argued that computational models of complex systems, like climate change, can capture useful information but can be difficult to apply to real-world scenarios of decision-making, such as those required in policymaking. This is because the type of information generated by large computational ensembles is often inconsistent with the traditional methodology used in decision analysis by policymakers. Lempert cited the more traditional methods of decision analysis as deductive (deductive arguments, for example), and newer methods as inductive (supercomputer power, for example). He argued that 'Computer-Assisted Reasoning' (CAR) was an approach to decision-making that was ideally suited under conditions of deep uncertainty because it allowed the application of complex systems to policy problems.

Decision analysis refers to the examination of a decision problem to ensure the decision model is the most appropriate and captures all available and relevant information at the time. In the context of policymaking in a liberal democracy, for instance, it refers to the fusion of the government branches of executive and legislature that is invariably dominated by the executive (see for example Maddison & Denniss, 2013). The 'executive' is the branch responsible for administering and enforcing the law. Moreover, the traditional 'rational-comprehensive' model of government decision-making is largely premised on 'review and recommendation', and debate and decision-making within the confines of the assembly (Lempert, 2002). Hence, it is, on the face of it, well removed from methods of statistical probability based on observation and experimentation.

Recent studies have advocated the use of robust decision-making (RDM) methods in response to the complex challenges posed by climate change mitigation and adaptation (Ramm, Watson, & White, 2018; Shortridge & Guikema, 2016; Tariq, Lempert, Riverson, Schwartz, & Berg, 2017). Climate change is characterised by deep uncertainties both in terms of real-world solutions, and the modelling that underpins the science. RDM is a quantitative, decision-analytic approach designed to augment decision-making under conditions of deep uncertainty. In short, wherein a consensus about the most appropriate model for a particularly complex decision-making problem is not apparent, RDM focusses on selecting the most robust options rather than the best options. The latter is more appropriate when future states are relatively predictable. When significant unpredictability is present – as in the case of climate change – implementing the best option(s) might entail significant risk.

To be sure, defining the criteria for what constitutes the 'best' option is a limiting factor. Moreover, RDM might not be economically feasible, and traditional decision-making methods might remain the best option in the case of data and resource constraints, and the nature of the climate change challenge, particularly as it applies to developing countries (Bhave, Conway, Dessai, & Stainforth, 2016; Dittrich, Wreford, & Moran, 2016). Furthermore, as argued by Todd Schenk (2018), the option of multiple scenarios accentuates the uncertainty complicit in the model, 'for better and worse'. Users are thus made aware of an uncertain future, but this does not inherently provide the means to make sense of its implications.

To account for the anomalies of unpredictability, robust decision-making focusses on options that meet the solution criteria across thousands of possible futures by introducing a broad spectrum of climate change scenarios (Guivarch, Lempert, & Trutnevyte, 2017; Lempert & Groves, 2010). Statistical 'scenario discovery' algorithms are used to recapitulate those futures wherein projected plans fail to meet expectations. This information provides a better understanding of the plans' weaknesses, and assists in eliminating those weaknesses. Arguably, as contended by Dittrich et al. (2016), robust decision-making methods provide the most potential to resolve issues of climate change uncertainty because they can be applied to a wide range of options with different degrees of complexity.

Central to the argument for robust strategies is that they are designed to perform well across a range of possible future climatic changes, and are therefore preferable to optimal strategies designed for one scenario that might perform poorly under a different, but equally probable, scenario. In light of the advent of climate change, its complexity and the uncertainties that confound its predictability, scenario approaches are becoming an integral part of robust decision-making (Guivarch et al., 2017). It is evident that the nature of the climate change problem gravitates around disparate mental models of future climate change scenarios; hence, the integration of a multitude of outcomes into a flexible model makes sense.

To enhance the robustness of computational models, Mayer et al. (2017) extended this concept to mental-modelling. Their aim was to more adequately account for the human reasoning process that underlies the decisions scientists make when confronted with trade-offs between different climate risk management strategies: the mental models that scientists form invariably inform computational models. Identifying the implicit preferences, value judgements, and biases not apparent in explicit models that underlie scientists' decision-making by eliciting scientists' mental models, would shed light on how they think about climate change and thus provide a better understanding of their modelling decisions.

To do this, Mayer et al. (2017) used values-informed mental models (ViMM), an 'augmented mental models research approach' developed by Bessette et al. (2017). ViMM provided an insight into the 'what and why' (what decisions were made and importantly, why they were made) of decision-making, of a group of 11 scientists. Although the group used a similar decision-making framework, results revealed that individual scientists hold quite different preferences, values, and beliefs in the course of modelling climate risk strategies. In particular, ethical and epistemic values were prominent in influencing modelling decisions: intergenerational equity was a concern for all scientists, along with the epistemic evaluation of the model's performance. This accounts for the value judgements – the human component – of scientists that drive their decision-making, and also elucidates the reasoning that underlies disparities between different model outcomes. Rather than applying the conventional deficit approach to mental-modelling, which ostensibly ignores values, this research provided insight into the epistemic and ethical values of a group with presumably, similar knowledge levels, compulsions, and objectives. By extension, of interest in terms of the application of ViMM, would be research into the epistemic and ethical values of a group of policymakers or politicians.

A principal point here was to draw a distinction between the traditional models of decision analysis used by policymakers, that are necessarily normative and therefore judgemental, and the models used by science that are descriptive and necessarily non-judgemental. Albeit humanly impossible to escape some degree of bias, the benefit of quantitative decision analysis (descriptive) is that it corrects fallacies in human reasoning (normative), and thus provides a systematic way to rank the appropriateness of alternative policy choices if it were to be applied to policymaking. However, the rationale of descriptive quantitation is often diametrically opposed to ethical considerations as a pretext to good public policymaking, and is therefore the defining factor that measures political action in terms of good judgement.

Extending this, robust empirical modelling that relies on large computational ensembles to fully capture a combination of probability and expert elicitation is often at odds with the traditional decision-making models used by policymakers. People have a general tendency to oversimplify the analyses of complex systems and the models they use, and routinely opt for intuition over scientific method and verifiable data (Northrop & Connor, 2013). Moreover, policymaking per se could be described as more a case of 'review and recommendation', as described by Lempert above, in which the number of recommendations to choose from can be inordinate. Added to this, in the presence of the political mayhem that has dogged the democratic world through the early part of the twenty-first century, getting a bill passed through congress or the senate, as the case may be, can be an excruciating task. Deliberative democracy appears to run second to plotting the politics of revenge.

From a purely pragmatic perspective, decisions should be based on rational and logical reasoning, should seek to maximise benefits and minimise costs over a predetermined timeframe while taking into account future accountabilities, remain focussed on the core issues, and push any distractions to the side. Suffice to say, endeavours to dehumanise decision-making to this extent has a zombie-like feel about it. Knowing that policy decision-making is exposed to, among other things, the peculiarities of the parliamentary process, party ideology and politicisation, public opinion and the populist vote, the quest for 'year on year' economic growth and jobs, the quest for environmental protection, social justice and equality, and the interjection of certain ideological worldviews espoused by the politico-elite and the politico-media complex. As political scientist Harold D. Lasswell (1960) declared, the paradoxes that pervade the social world cannot simply rely on logic as an adequate guide to policymaking.

As such, there is immense relief in being irrational and zombie-like. All at once the burden of great expectations is lifted. Similar in effect to a method of decision-making supposedly practised and perfected by the ancient Persians who would get exceedingly drunk in order to make an important decision with respect to predicting future outcomes. The decision was later reassessed in a sober state. If the same decision held sway in both a drunken and sober state, it was ratified. The theory being that one had to change one's mental and physical state to cast out uncertainty and come to a valid conclusion. In essence, will I feel the same in the future and in a different state of mind as I do today? Not dissimilar to Ernest Hemmingway's philosophy of writing drunk and editing sober.

Mike Hulme (2011) offered a more sedate perspective: 'The epistemological pathways offered by climate models and their derived analyses are only one way of believing what the future may hold' (p. 266). To be sure, a future reduced to endeavours to adapt to a changing climate was one way of envisioning the future. As Hulme (2011) puts it, greater endeavours should be made to include 'human creativity, imagination, and ingenuity' in the analysis of climate change prediction. Hulme challenged the idea of climate change determinism, suggesting that some of these futures might be better, some might be worse, but they should not be determined by the climate. Hulme's worldview could also be construed as another form of denialism: the anthropocentric mental model of climate change. Given the current state of the environment, one could reasonably argue that anthropocentric bias is likely to assign an overly optimistic probability to human ingenuity overcoming a climate catastrophe.

The IPCC (2013) stated that quantified measures of uncertainty in a finding are expressed probabilistically, and are based on statistical analysis of observations or model results, or expert judgement. The modelling parameters are agreed upon, and a consensus is reached based on these criteria. However, confounding this accord is something so contentiously complex that no operational definition exists to account for a lack of consensus as to which model is the most appropriate; in this instance, the 'particular decision-making problem becomes a case of deep uncertainty' (Institute of Medicine, 2013, p. 242). This is a common feature of parliamentary debate about climate change policy investments. Party ideology and politicking and the uncertainty inherent in the science become the foundation model for decision-making on a priori grounds; rather than substantive policymaking based on empirical evidence.

Deep uncertainty is the prevailing norm in politics. Even if expert advice is sought, exponents of the political art are fundamentally ill-equipped to deal with it. Holling, Gunderson, and Ludwig (2002) posed the question: 'why does expert advice so often create crisis and political deadlock? Why, in many places, does science have a bad name?' (p. 7). Can we comfortably presume that all those that profess to be experts are free from the tyranny of ideology or those that would impose their tyranny on them? That is, it would be remiss to give into the convention that political nuance has displaced expertise.

The question of whether the public was capable of constructively participating in policymaking was debated by Walter Lippmann (1922, 1927) and John

Dewey (1922, [1927] 2012). At the heart of the debate was the question of 'expertise' – the expert mental model as opposed to the lay mental model. Lippmann argued that it was unrealistic to expect the publics (non-scientists), as non-experts, to constructively contribute to the debate and design of complex policy. Lippmann's worldview was to leave complex matters to the experts. In contrast, Dewey was convinced that science was compatible with democracy and the publics' participation was fundamental to the democratic process. He argued that socioeconomic issues were the oversight of the public realm. Dewey's worldview was to ensure that the publics were substantially informed so they might knowledgeably and independently contribute to the scientific debate through the 'public sphere' – a sphere that transcended civil society and the State, and institutionally guaranteed the publics' democratic political participation – as envisaged by sociologist Jürgen Habermas ([1962] 1991). Public participation in the science debate, however, could also be viewed as the diminution of the authority of science in the policymaking process. Assuming, of course, that policymaking was based on scientific reality.

Nothing is certain but uncertainty (*rien n'est sûr que la chose incertaine*). Thus, has the uncertainty inherent in the science allowed science to be surreptitiously reframed? Does it no longer command the authority of truth? Has it been reduced to just another opinion in the political debate? I suspect that although evidenced-based policymaking stands as the normative principle, it is often a case of opinion masquerading as science. Politicians succeed in blurring the boundaries between anecdotal evidence and empirical evidence. It is acknowledged that the scientific method is not perfect. This was fairly atoned for by Charles Lindblom (1959) and his 'science of muddling through', in which policymaking decisions, in the absence of 'superhuman comprehensiveness', are made incrementally, or 'on the run'. He would later describe this as 'disjointed incrementalism', whereby policymakers suffer the constraints of inadequate knowledge and understanding. Therefore, 'no more than small or incremental steps – no more than muddling is ordinarily possible' (Lindblom, 1959, p. 517). However, from the publics' perspective, there appears to be no boundary between muddling and mayhem.

For the most part, disjointed incrementalism, or muddling, is a most apt description of the fragmented approach to global warming by nation-states. One might surmise that this persistent fragmentation serves the interests of the powerful (see for example Biermann, 2014). It affords them the flexibility to selectively modify their transnational agreements when they no longer prove convenient – CO_2 emissions reduction agreements, for example. U.S. President Donald Trump's decision to withdrawal from the 2015 Paris Climate Agreement in June 2017 (with the stated intent to renegotiate a better deal for the United States) on the grounds that it was not in his country's best interests, is a case in point. Sovereignty guarantees the continuation of this behaviour. Sovereignty legitimises the idiosyncratic actions of nation-states and their leaders that contradict global norms. When a world leader reneges on its pledged emissions reduction, it creates uncertainty with respect to the efficacy of global climate change policy per se, and inspires propagandists to invent new ways of framing the character of climate change in favour of those idiosyncratic actions.

A preface to framing theory

Framing someone for a crime they did not commit provides one with a very general understanding of a particular type of framing. In the fields of psychology and sociology there are two main types of framing: equivalence framing and emphasis framing (Liu & Scheufele, 2016). The former eludes to different presentations of 'otherwise identical information' to encourage a preferred interpretation and discourage another. The latter refers to the construction of 'different sets of information' to purposely overstate certain aspects of an issue while understating and/or omitting others. Arguably, politico-media framing contains elements of both; however, in the main, the type of framing relevant herein is 'emphasis framing'. I thus define the art of framing as to consciously or unconsciously sway an audience's preferences by understating, overstating, misstating, and/or omitting, particular points of reference in a communication to create a preferred impression which, therefore, has ethical implications. Moreover, the strength of a framing effect is a function of the strength of the impressionability and/or the conviction of an audience's preferences. How political and scientific communications are framed is, thus, of the utmost importance.

Framing climate change as a genuine threat, or a hoax, has been a preoccupation, to varying degrees, for most of the actors and institutions that appear in this story. In essence, although framed and reframed in myriad ways, climate change is, and perhaps always was, a human thinking and acting problem. Framing illuminates the way in which influence is exerted over human consciousness by the transfer of information (Entman, 1993).

Consider the framing of poverty for example. In a positive frame: The World Bank has reported that over the past few decades great gains have been made in reducing global poverty. In 1990 35 per cent the world's population lived in extreme poverty. Today that percentage has been reduced to about 10 per cent. This means that since 1990 over a billion people have been lifted out of poverty and now live well-nourished and well-educated lives in safe and secure environments. If this trend continues, the total eradication of poverty is a realistic possibility. As opposed to a more negative frame: The World Bank has reported that more than 760 million people now live in extreme poverty, surviving on less than $1.90 (international dollars) a day. And although over a billion people have been lifted out of poverty since 1990, this can be largely attributed to China's remarkable socioeconomic development since transitioning to a market-based economy in the late 1970s. What is more, according to UNICEF, extreme poverty claims the lives of more than 22,000 children under the age of five, every day of every week. That is more than 150,000 deaths each week and almost eight million largely preventable deaths each year. These children die quietly in the impoverished corners of the world far removed from the conscience of the material one.

Although both of the frames presented foregoing are true, according to reports, it shows that poverty can be framed in a more positive way or in a more negative way, depending on one's bias. Framing emphasises certain points, omits others, and plays on the emotions of the audience. The fact that 22,000 children under the age of five die from poverty every day, *shameful as it is*, cannot help but strike an emotional nerve. Likewise, communicators use emotions to both motivate and deter support for climate change action (Salama & Aboukoura, 2018). The emotional framing of an issue can have a profound impact on an audience; whether this creates a permanent change in attitude and behaviour, is another question.

Research by Hostler, Wood, and Armitage (2018) indicated that emotional cues that are positively valenced can enhance prospective memory performance. Prospective memory can be defined as remembering to perform an intended action in the future without an external prompt to do so. Prospective memory is therefore crucial to ensuring that one fulfils one's intentions. As such, in the context of framing, prospective memory is instrumental in modifying an audience's behaviour in accord with the sender's intent. I submit, however, that a comprehensive investigation of the role of emotions in reasoning is for another book. Even a concise definition of emotion is not easily had (Dub, 2017). It is, perhaps, best explained succinctly in the words of Richard Dub (2017): 'a life stripped of emotion would be a life stripped of meaning' (p. 245). To be sure, moments etched in one's memory often invoke emotions of both jubilation and despair, and of course, intent.

Memory is a fundamental cognitive capacity. Remembering is an imaginative process that entails the construction of a representation of a past event based on information currently available (Michaelian, 2017). Psychologist Wolfgang Schneider (2015) has argued that memory is at the heart of human cognition: few acts of cognition, whether basic-level or higher-level processing, do not involve memory. All decision-making and problem-solving processes depend on the retrieval of information stored in short-term (short-term store) and long-term (long-term store) memory systems. The former stores information long enough to be evaluated, perhaps only seconds, while the latter stores information indefinitely. Two memory storage systems are thus designed to perform two distinct functions. How we frame, store, and retrieve information is important going forward. Mindful that humans are susceptible to 'memory failure'.

Our memories are malleable and usually fade over time (Hostler et al., 2018). Psychologist Daniel L. Schacter (2001) referred to what he called 'the seven fundamental sins of memory malfunction': transience (memory diminishes or is lost over time); absent-mindedness (the interface between attention and memory is disrupted); blocking (the retrieval of information is temporarily thwarted); persistence (recalling memories that cannot be forgotten although there is an urge to do so); suggestibility (a memory is implanted by way of suggestion, comment, or a leading question); bias (the formidable influences of current knowledge and belief as they unconsciously rewrite the past); and misattribution (a memory is attributed to the wrong source and consequently mistaken for reality when it is fantasy or merely opinion). All 'seven sins' are relevant herein, and are particularly relevant in the course of framing public opinion. Public opinion is pivotal to this story.

Solon of Athens (Chief Magistrate in the city of ancient Athens in 590 BCE) was perhaps the first to formally recognise the value of popular consent among the publics. Like the creation and critique of journal articles one must do one's best to appease one's peers. But to avoid deviating from the subject of focus, public opinion is a function of information synthesised and assembled from an individual's memory. Information is framed by the sender, decoded by the receiver, and a mental model formed as a consequence. Extensive research has demonstrated that what the publics think about a particular issue is influenced by how it is framed (or portrayed) by the politico-media (see for example Bailey, Giangola, & Boykoff, 2014; Östman, 2014; Schlichting, 2013; Schuldt & Roh, 2014). In this way, one would naturally assume, that public opinion is a reflection, to some mutable extent, of pervasive politico-media framing. By extension, the media's importance as a source of information and conduit of communication between policymakers and the publics – the media are the principal gatekeepers of this conduit and are thus able to exert considerable influence on public opinion formation – means that both political actors and institutions can ill-afford not to treat the media with ample care and consequence (Esser & Strömbäck, 2014).

Media, the plural of medium, is anything that communicates information or data: from mobile media devices to our own voices and, of course, the Internet; the list is extensive and growing in complexity (Esser & Strömbäck, 2014). The politico-media complex (PMC) is that often incongruous but mutually beneficial relationship that exists between the Media and a state's politico-elite. My focus here is on the news media in particular, as a collective of sociotechnological institutions, and its association with the politico-elite – or ruling class. Actors and institutions of the ruling class can advance their respective interests by intentionally manipulating the framing of an issue. Political scientists Dennis Chong and James N. Druckman (2007a, 2010) have argued at length that citizens' attitudes can be substantially influenced by how elites – the ruling minority – choose to frame their communications through the news media. It is no secret that politico-elites and their political parties continually contest what will become an issue, and then carefully devise how to frame it.

For instance, if enough people buy into the frame that climate change mitigation and adaptation is secondary to economic growth and terrorism, then economic growth and terrorism will remain the overriding premise. Notwithstanding the paradox that the universal quest for 'year on year' economic growth, in concert with population growth, is the fundamental driver of climate change through CO_2 emissions emanating from fossil fuel combustion. This connection is rarely made and never publicised by politicians that are shackled to the frame of 'jobs and growth'. Connecting the dots requires deeper critical thinking. In the case of terrorism, the defeat of ISIS (Islamic State in Iraq and Syria) in the dying stages of 2017 came with little fanfare. It was almost like the politico-elite and the PMC were disappointed at losing an effective political ploy and media-grab (see for example Louw, 2010) to divert attention away from other contentious issues like climate change, inequality, and racial discrimination.

Keith E. Stanovich (2009), a Professor Emeritus of Applied Psychology and Human Development, described someone who is not predisposed to deeper and more systematic critical thinking as a 'cognitive miser'. Cognitive misers are people who have a limited capacity to process information. They have a tendency to take shortcuts that might lead to adequate outcomes but also deeply flawed ones. Most of us are more or less guilty of this when it comes to the routine decisions of the everyday. One fails to think fully because a shallower type of reasoning comes to mind so quickly. When exposed to framing, however, this flight to efficiency might allow whoever is doing the framing to literally make up another's mind for them. The sophisticated framer might have more control over the political and economic behaviour of an individual than does the individual. In this sense, framing effects are a genuine threat to personal autonomy and thus, have the potential to reframe the democratic principle. In essence, framing theory implies that the framing of a communication can influence an individual's thinking and acting to the extent that it undermines one's autonomy. It is therefore reasonable to suggest an association between framing and the mental models formed as a consequence.

In summary, frames emphasise certain pieces of information thereby giving them salience. A heightened salience enhances the possibility that the individuals in an audience will discern the meaning, process the information, and store it in their respective memories, thus prompting a certain mode of thinking and acting. According to philosopher and cognitive scientist David J. Chalmers (2010), a system is conscious of information when it has the capacity to act on that information; or more precisely, when it attends to, integrates, or exploits that information to affect a behaviour. It can therefore be said that the purpose of information is to modify a behaviour or reinforce an existing one. Of course, information might be framed disingenuously; and the purpose rendered immaterial if the system does not have the proclivity or the capacity to act on that information.

A preface to mental model theory

Have you found yourself in a place or situation that was not what you had in mind: that it does not match the mental model that you pictured in your head? Perhaps your reasoning process was flawed and you relied on too many ill-founded assumptions. There are two main theories of human reasoning: the 'mental logic view' and the 'mental model view' (Oaksford & Chater, 2009). The former is founded on a model of deductive reasoning that involves logical calculations based on formal rules of inference. The latter contends that human reasoning is a process of model validation based on the comprehension and degree of truth in premises (Johnson-Laird, 1983, 2013; Johnson-Laird & Byrne, 1991). In effect, the former stresses content-independent reasoning processes based on logic, as opposed to the latter which stresses content-dependent reasoning processes based on a syllogistic approach. The debate between these two worldviews continues but I shall not enter into it here (see Gellatly, 1989, for a detailed overview). I submit that the effect of the content of a problem (its complexity and contention, and how it is framed) and its relevance in syllogistic reasoning, is more attuned to solving real-world problems. I therefore subscribe to mental model theory and will explain why as this story unfolds.

A principal strength of mental-modelling (as a methodology to enhance human reasoning) is the capacity to compare mental models of a problem (climate change for example) replete with their multifarious complexities, misconceptions, and omissions, to an expert model of the reality of the science (Wood & Linkov, 2016). The aim is to determine what influences drive an individual's thinking and acting. Using questionnaires and other forms of evaluation, mental model diagrams can be constructed that shed light on how and what individuals think about a particular issue. These diagrams can then be compared, using quantitative metrics, across different stakeholder groups and individuals to identify synergies and antagonisms, important beliefs, and the levels of existing knowledge. Comparisons can reveal opportunities for interventions that lead to the formation of more correct, or competent, lay mental models.

Mental-modelling also has its limitations (Wood & Linkov, 2016). It can be difficult to implement wherein structural differences between stakeholder models exist, as they relate to beliefs, biases, and perceptions that result from cultural, social, organisational, and individual systems of thinking, and possibly the structure of one's brain. This is because the method used is commonly based on a deficit model: two archetypical mental models are compared, a layperson model to an expert model. In short, complete expert knowledge is compared to less complete layperson knowledge. The aim is to increase the layperson's knowledge to that of the expert's knowledge to enhance the reasoning process and thus the quality of decision-making. Mindful that there are different kinds of knowledge and corresponding intelligence (Sternberg, 2017). Depending on the circumstances, street smart might trump science smart, for example. As such, as structural differences increase, so does the difficulty to achieve an empirical comparison (Wood & Linkov, 2016). To resolve these issues, computational tools and algorithms to convert the elicited data into mental model diagrams are becoming more readily available. Of course, the source credibility and method of elicitation (accounting for bias, sample size, and interpretation) are common generic limitations. Nonetheless, mental-modelling presents a way to analyse and correct unsubstantiated reasoning and thus, irrational decision-making and behaviour.

The mental model theory was developed by psychologist Phillip Johnson-Laird and cognitive scientist Ruth M. J. Byrne. As defined by Johnson-Laird (1983), mental models are blueprints for thinking and acting. Since the early 1980s mental model theory has – arguably – become the most influential theory of human reasoning. There are now mental models proposed for almost every cognitive function. I define mental models as personally formed, internally held cognitive representations that have, to a mutable extent, an analogical coherence with the external physical world that affect a person's thinking and acting. Notwithstanding that gaps, or blind spots, may exist and values and beliefs may exert an influence. Of particular relevance here is that humans have difficulty forming a mental model that accurately simulates the scenario of climate change (Boschetti, 2012; Chen, 2011). This in turn affects their thinking and acting with respect to how they respond to climate change.

No doubt climate change embodies a complex set of challenges. Partly because it is manifest with risks that are not easily observed and identified – risks that humans have trouble assessing. The construction of human risk perception is a complex, multidimensional process; as such, a profound public understanding of the implications of climate change is perhaps beyond reasonable expectations. Further confounding this are the interests of the powerful and persuasive. If vested interests should embark on a deliberate ploy to confuse or manipulate the publics' understanding of climate change, and thus manifest a deficient or misleading mental model of climate change, the publics' conception and therefore their opinion of the climate change problem might become stranded in a false premise.

The probability of being stranded in a false premise is diminished by rational thinking, which is most relevant to this story. Rational thinking and decision-making can be linked to the concept of 'mindware' and crystallised intelligence ('a model for the assessment of rational thought'), as described by Toplak, West, and Stanovich (2014), Stanovich, West, and Toplak (2016), and West and Stanovich (2015).[7] Mindware, adopting the analogy that 'mind is to brain as software is to hardware', is much like computer software, comprising 'the rules, knowledge, procedures, and strategies that an individual retrieves from memory to aid in decision-making and problem solving' (Stanovich, 2012, p. 355); in effect, the knowledge bases that can be used to override intuitive responses. Not all mindware is conducive to rational thinking, however. Contaminated mindware can interfere with rational thinking because of mindware gaps that can lead to irrational decision-making and behaviour. Contaminated mindware occurs when an individual acquires one or more 'crystallised inhibitors' (West & Stanovich, 2015). Crystallised inhibitors consist of dysfunctional personal beliefs and superstitions, value placed on unsubstantiated knowledge, an overreliance on introspection, and unrealistic optimism. By contrast, 'crystallised facilitators' refers to declarative knowledge that can aid in rational thinking, comprising knowledge of scientific reasoning, financial and economic thinking, and statistics and probability. For instance, according to Stanovich (2009), a common mindware gap is lack of knowledge of probability theory, which is pervasive and the basis of much irrational thinking and acting.

It is apt to make the distinction here that contaminated mindware is associated with cognitive biases that might result in suboptimal thinking and acting; whereas suboptimal thinking and acting as a consequence of a paucity of knowledge of probability theory, for example, might be more to do with the capacity or opportunity to learn. This means that an individual who fails to learn or comprehend probability theory (which can be not easily grasped as explained later) that results in irrational thinking and acting could hardly be categorised as irrational on that basis. Ignorance is not solely a condition of choice. However, if knowledge of probability theory is a prerequisite for rational thinking and acting in certain circumstances, say policymaking or policy advocacy, then sub-optimal thinking and acting are explicably viewed as irrational. The distinction being, that to remain ignorant by choice when one has the opportunity and capacity to grasp probability theory is prescriptive of irrational thinking and acting, as opposed to the lack of opportunity or capacity to rectify that ignorance.

Further to the peculiarities of probability theory, the enterprising cognitive scientists Amos Tversky and Daniel Kahneman (1983), found that given even mild incentives individuals might succumb to the conjunction fallacy. The conjunction rule states that 'a conjunction cannot be more probable than one of its constituents'; that is, 'the probability of the intersection of events, or conjunction, cannot exceed the probabilities of its constituents' (Tversky & Kahneman, 1983, p. 293). It is safe to assume, however, as Tversky and Kahneman duly noted, that people do not normally evaluate the everyday in terms of probability theory or indulge in calculating the compound probabilities by aggregating rudimentary ones. Arguably, considering the longer-term implications of climate change, individuals are even less likely to pay attention to the aggregated probability of the advent of a new climate system; particularly one that is much less human-friendly than the current one. Moreover, not only is probability theory essential for predicting when or if this will occur, but of prime importance when endeavouring to navigate an increasingly complex world. I surmise that mental models constitute the internal compass for this navigation. Hence, a deficient mental model is a precursor to getting lost, both mentally and physically.

The concept of mindware has implications for mental model formation. It is proposed that a deficient or misleading mental model is a product of contaminated mindware or mindware gaps. A deficient mental model is thus the agent of irrational thinking and acting. To this end, I will borrow from the theories of mindware and rationality espoused by Stanovich (2009), Stanovich et al. (2016), and Toplak et al. (2014), and will refer to a suboptimal mental model, a product of mindware gaps or contaminated mindware, as a 'deficient mental model'. A deficient mental model, then, is one that has limited efficiency because it has acquired unsubstantiated knowledge that is mistaken for true belief. True belief, of course, is an ephemeral thing. It might not be true at all, even though at the time it was rational to think so.

Stanovich (2009) put forth the proposition that contaminated mindware could quickly spread through a population like an epidemic: a virus that corrupts the mental models of its unwitting users. Such a virus could be intentionally or unintentionally deployed in the course of 'social engineering' – a deliberate fine-tuning of the masses through mental model manipulation. For instance, have we been socially engineered to believe that to aggressively pursue clean energy and make environmental protection the priority will somehow result in economic ruin and mass job losses? Even though research suggests that the majority of people in advanced liberal democracies (Europe and the United States) were of the opinion that protecting the environment was more important than achieving economic growth (Drews, Antal, & van Den Bergh, 2018). Notably, having a job and having an opinion are two very different things. A better understanding of the process of mental model formation and the comparative disparities emanating from its counterpart – worldview – held by individuals and the publics, is indubitably needed.

The worldview concept as it relates to mental model theory and Dilthey's dilemma

Sustainability scientist Leo Elshof (2010) posited that in order to respond eloquently to the multifarious and interdependent problems of climate change one must first recognise that a person's perception of climate change is ingrained in his or her 'mindset or worldview'. Suffice to say that an array of disparate worldviews characterises contemporary culture. For mine, in his wonderful book, *Naming the Elephant: Worldview as a Concept* (2015), James Sire provided the most fitting introduction to the concept of worldview: 'For any of us to be fully conscious intellectually we should not only be able to detect the worldviews of others but be aware of our own – why it is ours and why in the light of so many options we think it is true' (p. 14). Of course, some discreet detective work is required in order to fathom the worldview of another person – and of course, some reflection.

A pause here is necessary with respect to 'reflection'. In accord with Dewey (1933), to reflect is to turn the thing over in one's mind, to seek additional evidence that will develop the proposition, and will either 'bear it out' or expose it as absurd and irrelevant. In terms of a reflectivist doctrine, which is germane to the autonomous human being, as portrayed by John M. Doris (2015): 'the exercise of human agency consists in judgement and behavior [*sic*] ordered by self-conscious reflection about what to think and do' (p. 19). For the purpose my story herein, more explicitly, reflection is a deeper more effortful mode of thinking and reasoning in which the fundamental goal is to validate a plausible conclusion and subsequently form an empirically accurate mental model that acts as a blueprint for thinking and acting. Importantly, a primary function of reflection is to augment one's worldview.

Worldview is probably the most important concept in the history of the world. Wars have been fought because of it. Planet Earth might even succumb to it. Change a person's worldview and you change his or her thinking and acting. In this story, the concept of worldview parallels that of mental model. They both describe how an individual relates to and interprets the world in the course of his or her thinking and acting in the everyday. I take this a step further and make the assertion that they are one and the same – worldview is simply the colloquial vernacular for mental model. Let me further explain why.

The term worldview is derived from the German *Weltanschauung*, meaning 'one's total outlook on life, society and its institutions' (Wolman, 1973, p. 406).

The celebrated Austrian neurologist Sigmund Freud ([1960] 2001), referred to *Weltanschauung* as 'an intellectual construction which solves all the problems of our existence uniformly on the basis of one overriding hypothesis, which, accordingly, leaves no question unanswered and in which everything that interests us finds its fixed place' (p. 158). Further to his critique, Freud ([1926] 2013) described *Weltanschauung* as 'Handbooks to Life' that soon grow out of date requiring new versions so that philosophers have information on every topic in order to continue their life's journey (p. [96] 20).

Freud was, of course, voicing his disapproval of philosophers and philosophy in general as an affront to the scientific method. However, he is at the same time endorsing the concept of worldview as central to our endeavours to understand the world. Perhaps ironically, Freud was also expressing his worldview in defence of his own ego as a neurologist and self-proclaimed scientist – his worldview was indubitably different to that of a philosopher's. A case, perhaps, of the interminable contestation between researchers' theory and theoreticians' theory, and an instance of how peoples with diverse backgrounds hold diametrically opposed views of the world.

Philosopher Stephen Pepper's ([1942] 1970) construct of a 'world hypothesis' parallels that of a mental model: a world hypothesis is a model of the universe composed of observations and inferences that individuals use to explain their reality. One could easily replace 'world hypothesis' with 'mental model' in this instance. According to Pepper, each world hypothesis is an endeavour to provide a complete view of the world; however, it has infinite scope and is sublimely accurate to the point that it will only validate one interpretation of an event. That is to say, each world hypothesis is autonomous and does not compete with other world hypotheses. But there is an air of absolutism about Pepper's construct – the competition among world hypotheses, or mental models, is what brings us closer to the truth. There might only be one truth, but finding it cannot be absolute.

Philosopher and poet Michael Boylan (2013) defined 'worldview' as 'a current personal consciousness that consists in one's understanding of the facts and about the values in the world. It is the most primitive term to describe our factual and normative conceptions' (p. 3). In the same way, mental-modelling, as we shall see, is fundamentally a competition between factual and normative conceptions. A point to note here is that research tells us that it is an individual's worldview, that for the most part, drives his or her perception of climate change, as opposed to his or her scientific understanding of it (Stevenson, Peterson, Bondell, Moore, & Carrier, 2014). One can thus reasonably ascertain that one's mental model of climate change does the same. Relevant here is that failures in an individual's worldview model – or mental model – will invariably lead to failures in that individual's rational thinking and subsequent decision-making with respect to climate change. Consequently, if enough individuals fall into this category, it stands to reason that climate change policy will likewise fail.

In summary, James Sire's (2015) elucidation of Wilhelm Dilthey's (1988) conception of worldview is most pertinent here: a worldview is a 'set of mental categories arising from deeply lived experience which essentially determines how a person understands, feels, and responds in action to what he or she perceives of the surrounding world and the riddles it presents' (p. 27). It is also interesting that Dilthey (1988), a historian, philosopher, psychologist, and sociologist, suggested that science will ultimately supplant worldviews, but did not elaborate as to how this would transpire. In light of the argument presented foregoing, I submit that mental model theory, and the science of mental-modelling, represent the solution to Dilthey's dilemma.

Nature versus nurture: the blessing and the curse

Historically, humankind's place in the world was confounded by two principal worldviews – nature versus nurture. Contemporary thinking suggests that a person's worldview is one of choice or is passively accepted as a condition of the culture into which one is born. Debate, however, continues around which holds sway – nature versus nurture – as the principle predictor of, and influencer on, human performance and behaviour. By definition, Charles R. Darwin's theory of natural evolution endorses the worldview of human evolution by natural selection. Natural selection involves the accidental and incremental transmission of genes to the next generation with the enterprise of maximising an individual's life chances through adaptation to a particular environment. There are no guarantees of an optimal outcome, however.

There is broad acceptance of the 'selfish gene' which refers to the entity that succeeds as a consequence of natural selection (Dawkins, 2006b). Naturally, the selfish gene is the one that outdoes all others and replicates to ensure the 'survival of the fittest'. Strictly speaking, a gene is 'the smallest discrete unit that is inherited by offspring intact' (Buss, 2008, p. 10). Broadly speaking, it is a single unbroken sequence in our DNA codebook that determines the characteristics of our mind and our body. Most importantly, genes are the universal transmitters of powerful adaptative knowledge. 'Fitness' refers to an organism's proficiency at transmitting the genes that prove most advantageous in the quest for survival.

This does not imply genetic determinism, however, as this would suggest that human behaviour is controlled by genes alone. Evolutionary theory requires two requisites: (1) Nature: adaptation through evolution and natural selection (2) Nurture: environmental input that triggers the development and activation of this adaptation (Buss, 2008). That is, determinism, as opposed to free will, suggests that one's life experiences combine with variants of particular genes to influence behaviour, but only if those genes are susceptible and accommodating. Moreover, in accord with Darwin's theory of natural evolution, the brain evolves in concert with human intelligence – the advent of grey matter – that in turn determines its capacity to organise human thinking and reasoning.

What distinguishes thinking from reasoning? David Moshman (2013) defined thinking as 'the deliberate application and coordination of one's inference to serve one's purpose' (p. 15). This encompassed an array of interrelated cognitive processes such as 'problem solving, decision-making, judgement, and planning'.

Whereas, he defined reasoning as 'epistemologically self-constrained thinking – that is, thinking aimed at reaching justifiable conclusions' (p. 15). This encompassed 'logical reasoning, scientific reasoning, moral reasoning, and argumentation'. A key point of difference is intent. The goal of thinking is 'to serve one's purpose'. The goal of reasoning is to reach a justifiable conclusion. Which might serve one's purpose, or not. Thus thinking, as opposed to reasoning, has profound implications for one's worldview.

In accord with renowned sociologist Anthony Giddens (1994, [1991] 2013), worldview is a consequence of social evolution and is nurtured and thus shaped by the society into which one is born. Or as journalist Robert Wright (1995) put it: the malleable mind begins as an empty flask that is progressively filled with the surrounding culture that dissolves its evolutionary roots. Humans are thus a product, in the main, of their cultural environments, peers, and parents (Lieberman, 2016). Normative social influence refers to the pressure exerted by others to conform to their expectations. That said, divergence from the norm often occurs as individuals trust in their own learning, and dissent from the majority to form their own beliefs and opinions (Hodges, 2017) – a case of thinking and reasoning for oneself. The traditional faculty of socialisation has also been tempered somewhat by a post-traditional society that does less to socialise individuals as a consequence of its globalising proclivities (Wright, 1995). Subsequently, from the conventional nurture perspective, reasoning might have little to do with one's worldview.

Nurture is most often not by choice. The revolutionary moral and political philosopher Karl Marx, in his 1859 Preface to *A Contribution to the Critique of Political Economy*, observed that men, in the social production of their existence, were destined to be entangled in a hierarchy of relations that were beyond their control. These relations were defined as the capitalists and the proletariat (or workers). The entirety of which constitutes 'the economic structure of society' on which the 'legal and political superstructure' is built (as cited in Rigby, 1987, p. 187). This superstructure corresponds to an individual's social consciousness as the creation of the Social Contract he or she inevitably enters into.

From the social evolution perspective, as a general rule, one's worldview is an outcome of the company one keeps and the place in which one lives and works. Mindful, however, that research also tells us that 'children and adults demonstrate both selectivity and fidelity in following others' actions' (Hodges, 2017, p. 100). Moreover, in the present-day, the media individuals consume must be considered as playing a role in their social evolution. For instance, philosopher and author Alain de Botton (2014) declared that the news 'does not merely report on the world, but is instead constantly at work crafting a new planet in our minds in line with its own often highly distinctive priorities' (p. 11). He further claimed that endeavours in the classroom had negligible bearing henceforth as the most compelling and ongoing tutelage took place on our television screens and through other digital technologies. In this sense, the politico-media framing of climate change could be considered a modern form of nurture – even to the extent of social engineering a preferred attitude and behaviour.

In contrast, one might view mental-modelling as an evolutionary adaptation, the incremental cognitive capacity to accurately mental-model the future. For example, theoretical physicist Michio Kaku (2014) has suggested that the evolution of human consciousness has allowed humans to construct a model of the world and then project that model into the future in order to realise a goal. Indeed, the nature versus nurture debate might be unwinnable from a Marxism/Darwinism worldview; however, nature (mental-modelling) versus nurture (framing) as possible contributors to the success or failure of climate change policy, is central to the plot herein.

It is pertinent to add here that the debate over nature versus nurture is a long-standing one. It has gone through a dialectical cycle of changes (Dai, 2017). Initially, nature and nurture were treated as separate entities, as additive contributors to the development of superior performance. The second stage of the cycle reconsidered them as reciprocal, each one amplifying the other. The third stage viewed nature and nurture as inseparable through the human developmental process: 'nature reveals itself through nurture, and nurture mediates nature (e.g., genetic expression)' (Dai, 2017, p. 5). Henceforth, biological and social factors can be linked to a single theory (Fox, 2017). To reconcile the role(s) of nature and nurture, however, as David Yun Dai contended with respect to a theory of giftedness (endowed with some form of superior natural ability), required a marshalling of all the evidence. And there is a large volume of evidence that points to natural endowment as a significant factor in superior human performance.

Certainly, in the first instance, one would presume that a superior mental-modeller is one who is naturally gifted. The idea of being 'naturally gifted' tends to resonate with most observers: the mind-boggling thought experiments of Albert Einstein, the extraordinary athletic exploits of Jesse Owens and his contemporary Usain Bolt, the sublime musical compositions of Ludwig van Beethoven, the literary genius of William Shakespeare, and the fabulous wit of Oscar Wilde, to name but a minuscule of the many. Moreover, Macnamara, Hambrick, and Oswald (2014) conducted a meta-analysis which showed that practice (nurture) was not nearly as important as natural talent (nature). They found that deliberate practice relative to performance explained 26 per cent of the variance for games; 21 per cent for music; 18 per cent for sports; 4 per cent for education; and less than 1 per cent for professions (p. 1608). It seems, therefore, practice can only make perfect if nature has first provided an adequate measure of the fundamentals to achieving as much.

Notes

1 SAD is categorised as an 'unspecified depression disorder' by the American Psychiatric Association because the criteria needed to make a more specific diagnosis cannot be met. A point to note is that SAD is not associated with psychosocial stressors, such as regular unemployment at a particular seasonal time of the year, for example.

2 'The Earth's energy imbalance is the difference between the amount of solar energy absorbed by the Earth and the amount of energy it radiates back to space as heat'

(Farmer & Cook, 2013, p. 81). If the imbalance is negative then the Earth will cool, as more energy is going out than coming in. The Earth now retains more of the electromagnetic radiation from the sun than it radiates back to space. This energy imbalance is the single most important measure of the Earth's climate as it explains predictions of future climate change. Radiative forcing (RF) quantifies the change in energy fluxes caused by the natural and anthropogenic drivers of climate change. Positive RF results in surface warming while negative RF results in surface cooling.

3 Albedo is a measure of the reflective qualities of a body defined by the ratio of reflected radiation to incoming radiation. For example, freshly fallen white snow might have an albedo as high as 90 per cent while a black surface such as asphalt might have an albedo as low as 10 per cent.

4 See Rahmstorf (2009) for comments as to how realistic the 2004 disaster movie *The Day After Tomorrow* is in terms of climate change modelling.

5 Hanna et al. (2013) defined 'mass balance' 'as the net result of mass gains (primarily snow accumulation) and mass losses (primarily meltwater runoff and solid ice dynamical discharge across the grounding line)' (p. 51). Furthermore, surface mass balance (SMB), which can vary significantly, is the net balance of accumulation and losses at the ice sheet surface and does not include dynamical mass loss.

6 The terms and definitions used to classify the three types of 'uncertainty' by the 'Committee on Decision Making Under Uncertainty' in their report *Environmental Decisions in the Face of Uncertainty* (2013) were adopted for the reason that they were more appropriate and specific for the purpose of this story. Traditionally, aleatory uncertainty (randomness or natural variability) which is associated with seemingly random variation in natural or anthropogenic processes – things that are ostensibly outside our measuring capacity, and epistemic uncertainty (attributed to the lack of knowledge about biogeophysical processes) are more commonly discussed (see for example Beven, 2012, 2013; Deser, Phillips, Bourdette, & Teng, 2012).

7 The concept of crystallised intelligence can be traced to the Cattell–Horn–Carroll (CHC) theory of cognitive abilities (see Flanagan & Dixon, 2014). CHC theory represents the combined works of Raymond Cattell, John Horn, and John Carroll. It is recognised as the most comprehensive and empirically validated psychometric theory of how cognitive abilities are structured. Many of the methods used today to test intelligence and cognitive abilities derive from CHC theory. The West and Stanovich 'model for the assessment of rational thought' referred to above, derives, in part, from the CHC theory of intelligence (see West & Stanovich, 2015).

2 The stealthy art of framing

Framing mental models

> Words can be relied on only if one is sure that their function is to reveal and not to conceal.
>
> (Hannah Arendt, *On Violence* (1969, p. 66))

Only by our actions shall the truth be revealed, as political philosopher Hannah Arendt (1906–1975) implied above. Likewise the aphorism, the road to hell is paved with good intentions. Theoretically, what one knows to be true – the cues that one uses to orient oneself to the world, one's worldview or mental model – should be reflected in his or her response to any given scenario. Humankind's response to what is known about the scientific reality of climate change has been thus far erratic and duplicitous. While the science is robust the climate change 'idea' remains ostensibly esoteric. To counter this abstractness, the 'catastrophe' representation (mental model) of climate change is commonly propagated to make the invisible visible. But like a murder without a body, a catastrophe without physical substance remains somewhat of a mystery and thus, can be more easily framed in that context. That is, knowledge of causality derives from our interactions with it, and climate change is not so amenable to this condition. Yet the grounds for denial are increasingly flimsy while the urgency rises in concert with the frequency and intensity of extreme climate events. In consideration of this, it is proposed that two main constructs either aid or impede successful climate change policy: politico-media framing (nurture) and the mental models formed (nature).

Framing theory

Contemporary framing theory has elements of the original 'frame problem'. As it relates to artificial intelligence (AI), the problem of how to 'connect only the relevant dots'. The frame problem was presented by John McCarthy and Patrick J. Hayes (1969) in their quest to fathom the artificial intelligence (AI) conundrum of how to integrate a formal logic-based representation of events, with the associated intuitively obvious non-events. In effect, the logic (reasoning as a logical process) versus inference (reasoning as an intuitive process) conundrum.

The challenge, when performing an action, is to limit the scope of reasoning to all of those changes associated with that action, and dismiss all others. Mindful that humans are not entirely adept at this either. To do this adeptly, one has to rely on one's relevance updating mechanism as the action is being performed. The significance of which is to make the correct decisions based on what is relevant as a scenario unfolds, without having to explicitly consider what is irrelevant. So, the frame problem, in short, as Jerry Fodor (2001) later described it, is about defining what information is relevant to what: figuring out what is useful and discarding what is useless.

Framing theory, in subtle contrast, is about emphasising only those bits of information deemed as relevant by the sender, in order to achieve a preferred action, or inaction as the case might be. Framing theory, from the receiver's perspective, is more attuned to Fodor's postulate: figuring out what is useful, discarding what is useless, and making the correct decisions both during and post the action.

Framing theory is premised on the supposition that an issue can be presented by the sender in a way that is likely to lead the receiver to the sender's predetermined point of view. Chong and Druckman (2007b) described framing as the process by which people develop a particular conceptualisation of an issue or organise their thinking, attitudes, beliefs, and ultimately their actions. According to Robert Entman (1993), framing is about selection and salience:

> To frame [from the framer's perspective] is to select some aspects of a perceived reality and make them more salient in a communicating text, in such a way as to promote a particular problem definition, causal interpretation, moral evaluation, and/or treatment recommendation for the item prescribed.
> (p. 52)

A third type of framing of sorts (as distinct from equivalence framing and emphasis framing) is referred to as 'agenda setting', particularly in a politico-media context (Scheufele & Tewksbury, 2006). Agenda setting theory explains how the mass media shapes the thoughts of an audience by giving some stories precedence while downplaying or discarding others (see for example Lei & Vargo, 2017; Princen, 2018). The aim is to develop a sense of what issues are important through priming audiences to assess politico-elites and their governments based on performance benchmarks set in accordance with those issues (Glynn, Herbst, Lindeman, O'Keefe, & Shapiro, 2016; McCombs, 2014; Scheufele & Tewksbury, 2006). Priming, itself, is a form of influence that operates indirectly by cuing audiences to focus on certain themes. This shapes the way audiences perceive and make judgements about issues linked to those themes and the different political figures involved. Of course, the union, not always amicable, of politics and media through the politico-media complex tends to blur the distinction between what might be construed as equivalence framing, emphasis framing, and agenda setting (see for example Sevenans, 2017; van der Pas, van der Brug, & Vliegenthart, 2017). A compilation of all three is most probably at work.

Framing can be further distinguished on the basis of its use of critical reasoning – a matter of argumentation versus persuasion. A sound argument is supported by the logic and reasons as to why the premises are true (Vaughn, 2015). To persuade, which is more germane to framing, is to influence an audience's opinion by any means available. This might include playing on an audience's emotions through rhetorical guile such as the use of sophisticated language and persuasive metaphors, powerful imagery, deception, threats, and propaganda, for example. Moreover, the strength of a framing effect is also a function of the strength of the impressionability and the conviction of an audience's preferences. An individual's preferences might be deeply inculcated or equally changeable.

Framing is one of the most important concepts in the study of public opinion (Chong & Druckman, 2007a; Druckman, 2001b; McCombs, 2014). Framing effects occur when two logically equivalent statements of a problem – framed in positive or negative ways to either understate or overstate selective points – lead individuals to make different decisions and judgements. Where people maintain preferences that are not sufficiently robust, framing might be used to modify those preferences. The framing of a communication can, therefore, change a person's attitude toward an object or issue by changing the relative weights given to competing considerations about that object or issue. And the more contentious the issue, the more elaborate the framing becomes.

Tversky and Kahneman (1981) argued that rational choice rests on the theory that one's preferences should not change in accord with changes in the way a scenario is framed. Here we are ostensibly referring to the 'rational person' – someone who adheres to logic and is largely immune to emotional distraction: much in the mould of *Star Trek*'s Spock, for instance. Spock, as most of us knew him, was the fictional philosopher played by Leonard Nimoy. Half Vulcan half human, Spock was defined by his fabulously logical attributes – and frequently reiterated as much: 'May I say that I have not thoroughly enjoyed serving with humans? I find their illogic and foolish emotions a constant irritant.'[1]

Nonetheless, humans are predisposed to be illogical and irritant – as a consequence, in the main, of biases and emotions. This also makes them susceptible to framing. To demonstrate how human preferences could be manipulated through framing effects, Kahneman and Tversky (1984) conducted the 'frequently cited', but nevertheless fundamental, experiment referred to as the 'Asian Disease Problem':

> Imagine that the U.S. is preparing for the outbreak of an unusual Asian disease, which is expected to kill 600 people. Two alternative programs to combat the disease have been proposed. Assume that the exact scientific estimates of the consequences of the programs are as follows: If Program A is adopted, 200 people will be saved. If Program B is adopted, there is a one-third probability that 600 people will be saved and a two-thirds probability that no people will be saved. Which of the two programs would you favor [*sic*]?
>
> (p. 343)

In this experiment, 72 per cent of respondents chose Program A. For the second part of the experiment, the same options were offered but framed in a different way: if Program C is chosen, '400 people will die'; if program D is chosen, 'there is a one-third probability that nobody will die and two-thirds probability that 600 people will die'. Adopting the alternative frame resulted in the percentage distribution being close to the reverse, with 22 per cent choosing Program C and 78 per cent choosing Program D. Kahneman and Tversky's experiment demonstrated that sophisticated framing can play a significant role in the manipulation of decision choices. The extent to which the percentage distribution changed from the initial frame (Programs A and B) to the alternative frame (Programs C and D) supported this hypothesis. This was, however, a controlled experiment in a simulated environment – a scenario discrete from the norm of the everyday. As such, and the reason for recording it here, for the most part, is to note the validity, or lack thereof, of such an experiment later in this story.

Chong and Druckman (2007a) pointed out that if citizens' preferences reflect no more than arbitrary changes in frames, the existence of framing effects prompts serious questions about the appropriate role of the publics, and thus public opinion, in the design and implementation of public policy; not to mention the potential to undermine the democratic system. Many commentators consider public opinion to be a cornerstone of democracy. That is, the publics' political preferences form the substance of democratic governance. The extent to which politico-elites influence public opinion through sophisticated framing effects is thus a question that requires an answer. There is one more point to make with regard to robust preferences – they are good if they are expressions of an accurate mental model and bad if they are expressions of a deficient one. In reference to the latter, an unshakeable conviction can be as deleterious as one that is easily swayed.

Media effects and politricks

Journalist James Painter (2013) declared that the media's role in communicating climate change is crucial – most people get their information about scientific phenomena from the media. How the media frames climate change provides a powerful interplay between an everyday understanding and the ways it is discussed by politicians, scientists, and the publics (Boykoff, 2011). It is evident that the content, tone, and consistency of a communication have considerable bearing on effective knowledge transfer and the construction of one's personal reality. More broadly, Austrian writer and critic Karl Kraus (1874–1936), as quoted in Ross (2009), deemed any 'misuse of language as an attack on morality and the basic dignity of humanity' (p. 158). Kraus insisted that much of the evil that plagued society, such as poverty and crime, could be traced to the corruption of language. He cautioned that it was the corruption of language by the press that caused the devastation of World War I, the Nazi reign of terror, and Hitler's ascent to power. The 'power of the press' has powerful implications.

Today the mass media has evolved into something with an almost immeasurable reach and immense audience. Wolfgang Donsbach and Michael W. Traugott (2008) cited two technologies that drastically altered contemporary politics more than anything else: television and public opinion research. Television commercialised politics across all sectors of society, made politics more personal, altered the behaviour of politicians and the skills they required, and ultimately led some researchers to claim that television had usurped traditional political institutions as the hub of political power. And of course, as a measure of the political actor's performance, 'the polls' dominate television content during an election.

In the present day, particularly across the Western world, the Internet has established itself as the modern-day agora of information exchange, and the essential platform for political actors to interact with citizens in endeavours to gain their endorsement. The latent capacity of digital media has incited widespread interest among politicians and their cohort of professional media personnel. Political leaders and other politico-mediaists use text interfaces, such as online social networking and microblogging services like Twitter, to express their personal and political views in endeavours to 'set the agenda' and influence public opinion with an aim to achieving popular consent (Lee & Xu, 2017; Pal & Gonawela, 2017). Suffice to say, U.S. President Donald Trump has made Twitter his preferred means of communication with the masses.

From the early 1900s, journalist and political commentator Walter Lippmann (1922) brought attention to the conscious adoption by politicians of sophisticated framing techniques complicit with the power of mass media to 'manufacture consent'. In endorsing this sentiment, Edward Herman and Noam Chomsky (1994) argued that the elite fix the premises of discourse and the interpretation by filtering out what is deemed unnewsworthy in accord with their economic, social, and political agendas. They considered this to represent the 'propaganda model' in which elites set the agenda, determine the tone and emphasis, and thus frame the issues. The propaganda model moves the spotlight from the message to the mechanism. The politician uses the mechanism – political 'sleight of hand' – to shift the publics' focus away from one issue and toward another. A carefully orchestrated 'politrick' purposefully designed to distract and delude while endeavouring to modify citizens' values and attitudes, and thus their behaviour.

Perhaps the greatest 'politrick' ever played by the politico-elite was to convince citizens that they live in a liberal democracy. Marxists referred to this as 'false consciousness'. People believe that they are autonomous beings; that they are conscious of their thinking and acting. However, their thoughts and actions have been duplicitously programmed through the politico-media complex – the propaganda model as described by Herman and Chomsky foregoing. The socialisation of the masses, engineered and propagated with capitalist values, is the grand ideal. A capitalist class ideology now pervades the everyday – 'let's get the economy moving, go out and buy something'. The perennial broadcast of 'jobs and growth' by the politico-media has successfully instituted a voracious culture of consumerism that serves to fill the coffers of the politico-elite. How rare it is to navigate the everyday without being relentlessly bombarded by advertising to purchase something. Suffice to say, the now universal economic growth model – continuous 'year on year' economic growth – is the main contributor to climate change, as well as human exploitation and slavery and other modern-day social ills such as momophobia and Internet Addiction Disorder.

Maxwell McCombs (2014) argued that newspapers, inclusive of their digital cousins, rely on a multitude of cues about the comparative salience of the topics on the day's agenda – from the lead story on page one to the repeated coverage of a story 'day after day'. The publics use these cues to decide which issues warrant the most importance. The public agenda, according to McCombs is, therefore, set by the news media that ultimately shapes public opinion. Former *New York Times* executive Max Frankel (1999) expounded the extent of his newspaper's agenda-setting influence:

> It is the 'house organ' of the smartest, most talented, and most influential Americans at the height of American power. And while its editorial opinions or the views of individual columnists and critics can be despised or dismissed, the paper's daily package of news cannot. It frames the intellectual and emotional agenda of serious Americans.
>
> (p. 415)

A media empire is often controlled by a media mogul. Rupert Murdoch is an exemplary of the modern-day media mogul. Author and former journalist David McKnight (2012) described Rupert Murdoch as a 'ruthless man, feared by politicians and envied by his business rivals' (p. 1). McKnight placed Murdoch – the Executive Chairman and former CEO of News Corporation, the dominant media organisation worldwide – at the forefront of a dangerous abuse of media power. According to political theorist John Keane (2013), most observers viewed Rupert Murdoch as one of the most powerful political figures on Earth. For this story, Rupert Murdoch is the archetype of the politico-elite.

McKnight (2012) claimed that Murdoch, a right-wing populist and climate change denialist, along with his executives, had intimate political links – in reference to the Labour and Conservative governments in the United Kingdom pre-2011 – to the extent that they had gained access to the cabinet and intervened in policy matters. In Australia, Murdoch's influence was borne out in a succession of campaigns (in 2010 and 2011) to undermine the legitimacy of the Australian Labor Government (progressive left). This was seen as a reprisal for the previous Labor government's stimulus package in the wake of the 2007–2008 global financial crisis (GFC), which ran counter to Murdoch's neoliberal ideology. Indeed, Rupert Murdoch's personal worldview is powerful and far-reaching and as such, has implications for democracy.

It is generally perceived that politico-elites are driven by self-interest in their quest to control the masses (Herman & Chomsky, 1994; Lebow, 2018; Meyer & Hinchman, 2002). This derives from the premise that democracy survives on the legitimacy of continued public support (Glynn et al., 2016). This is a fact of modern democratic governance as well as a historical one. Unless ruled by violence, all regimes require an ample measure of popular consent (Herodotus, 440 BCE). Politicians thus have a vested interest in maintaining a presence on the media stage and subsequently aim to be highly 'mediagenic'. As a consequence, they feel pressured into stage-managing their appearances to gain greater access to media coverage. To stage-manage is to manipulate one's performance as an actor would in a stage play. By doing so, however, politicians progressively commit to playing by the media's rules. Thomas Meyer and Lewis P. Hinchman (2002) referred to this as the 'process of colonisation'. The media colonises politics, and the politicians, governed by self-interest, are subsequently made accomplices to the transaction.

I would add here that although 'self-interest' infers a negative connotation, I ask that we remain neutral with respect to this concept for now. I would also add that a reasonable argument could be mounted to suggest that politicians have minimal control over how they are portrayed by the media and, as a consequence, have no choice but to bend to the media and its logic. Undoubtedly, there exists spirited debate between journalists and political actors over what issues and frames should be the emphasis of public attention. The use of politricks (political trickery) and attention-grabbing sensationalism, however, risk diverting the publics' attention away from substantive issues (Matthes, 2012; Strömbäck & Van Aelst, 2013). The desire to be seen and heard might mean that one or both parties relinquish some control over their intended message, and this might be a price worth paying if it means increased media exposure and viewer audience, but is hardly in the publics' interest.

Adding to the ambiguity of politico-media communications, Dietram A. Scheufele (1999) reasoned that media effects are constrained by the interaction of the mass media with its audience. The strength and comprehension of a media message are contingent on the education, beliefs, attitudes, and political orientation of the individuals that make up the audience. Considering this, the mass media message might be received in an unintended way. The communication itself is only part of the process by which individuals construct meaning (Gamson & Modigliani, 1989). Knowing whom one is addressing is of primary importance. One's life experiences, social interactions, and physical disposition are engaged in the process of constructing meaning. Consequently, the presence of frames in the communication process does not guarantee their influence over an audience's thinking and acting (Entman, 1993). An audience might hold certain substantive beliefs, ideological or religious, that are beyond politricks and the influence of sophisticated framing techniques.

A person's temporal disposition also influences the effectiveness of a framing effect. Situational contextuality refers to the concerns and influences of both the encoder and decoder of a communication in accord with the place and time of their particular setting (Boholm & Corvellec, 2014). In short, it shapes how encoders and decoders interpret words, metaphors, intentions, and priorities. How individuals make sense of 'intentions, word choices, information, and commitments' is contingent on contextual variables. For instance, irrespective of how well the sales pitch is framed, it is reasonable to presume that a person abandoned in the desert under a blistering sun and dying of thirst is more likely to choose a pot of water over a pot of gold. Similarly, the context in which political actors and citizens perceive and pursue their interests has considerable bearing on the meaning of a communication. The encoder and decoder (sender and receiver) play their respective roles in the manipulation of preferences. Variables as wide apart as empirical knowledge and superstition, place and time, influence the strength and meaning of a framing effect.

Framing public opinion

An early American study by Benjamin Page, Robert Shapiro, and Glenn Dempsey (1987) tracked changes in public opinion towards a number of public policy issues between 1969 and 1983. They focussed on cases in which policy advocacy for an issue changed within a short time period. They then analysed television news coverage of these issues over the same period (major news networks were the most influential at the time of the study). For each news story they carefully coded the source: from policy experts, to journalists reporting on the issue, to the President of the United States. Their analysis showed that the tone of the actors' statements was influential in predicting opinion change. The more positive the statements were, or advocacy for a particular policy option was, the more public sentiment was likely to move in that direction. Of significance, however, was who made the statement. Statements by news commentators and television presenters, and experts on the topic of discussion, produced larger shifts in opinion than did statements by politicians and special interest groups. This suggested that issues framed in a positive tone by people considered more knowledgeable and less biased by the publics were more likely to shift public opinion. In short, source credibility was instrumental in the formation and strength of public opinion. Although, arguably expected, an important finding that I shall revisit later in this story.

R. Douglas Arnold (1990) contended that a politician's foremost goal was re-election. It is a matter of weighing up the electoral consequences of being associated with a particular policy option and predicting current and potential policy preferences in terms of constituents – but this was 'more art than science'. He maintained that electoral outcomes rest 'partly on what citizens will allow, partly on what coalition leaders [party leaders] propose, partly on what strategies coalition leaders and opposition leaders adopt, and partly on what legislators [politicians] themselves prefer' (p. 119). The 'art' resides in the capacity of politicians to correctly read the opinions of their peers and constituents, and position themselves accordingly.

In a Westminster-style democracy, it is presumed that public opinion remains the foremost driver of public policy. One might also assume that democracy makes the normative assumption that majoritarianism equates to wisdom – or somehow 'ought to'. The manipulation of public opinion, however, confounds this assumption and thus democracy. A study by political scientist Tetsuya Matsubayashi (2013), using American National Election Study data (1956–2004) of how elected officials influence the opinions of citizens in which political issues are open to public debate, showed that incumbents did indeed move their constituents' preferences in a particular direction. The most effective representatives were those who received the most media exposure. This contradicted more traditional studies of political representation that indicated that citizens influence the positions taken by their politicians. One might reasonably presume, however, that politicians and the publics manipulate the preferences of one another in the course of public debate and the public policy process.

Matsubayashi (2013) concluded that politicians did actively seek to frame public opinion. He found that politicians not only seek re-election but also endeavour to institute policy preferences in the direction of party policy, along with their own preferred manifestos. This endeavour is made easier when politicians shift constituents' preferences closer to their favoured positions, which in turn allows them to pursue their own policy goals without suffering any electoral downside. The overall objective was to maximise their chance of winning elections while instituting their preferred policies. This is similar in effect to Arnold's (1990) hypothesis, however, it extends to include the self-interest of achieving both re-election and realising their own policy preferences as equally dominant goals.

Matsubayashi (2013) also found that 'highly persuasive' political rhetoric did not automatically guarantee that the politico-elite were capable of arbitrarily manipulating public opinion. Indeed, people with robust opinions or policy preferences might not succumb to the politico-elite's message. Moreover, there is a deeper question here. Is it safe to assume that the policy preferences held by political parties are based firmly on empirical evidence? Given the potential for politicians to influence public opinion through sophisticated framing effects, would it be unethical to propagate policy based predominantly on party ideology? Likewise, making a policy promise knowing that it would be very difficult or near impossible to fulfil. That said, when the chant goes up, 'show me the evidence' – that climate change is of no real concern and is not human-induced and does not require urgent attention – politico-elites are adept at digging up a contrarian scientist intent on fabricating the frames of uncertainty (Farrell, 2015). Not in the name of science but in the name of their neoliberal brethren. Suffice to say, climate change policy must be founded on empirical evidence, and not the debased political sloganeering which has become the norm. Moreover, what is the essence of a democracy if public opinion is not respected and safeguarded?

Framing climate change

In this story, the Intergovernmental Panel on Climate Change (IPCC) is recognised as 'the' epistemic authority on climate change policy and, therefore, the 'expert mental model'. No doubt there are those that would disagree with this. Beck et al. (2014) have argued that the IPCC's framing of climate change as a 'universal global risk' is contributing to political inertia. They contended that this very specific framing of the climate change problem, which focusses on achieving a global consensus, has divided the publics and rendered climate change policy ineffective. By focussing on a global consensus to support policy decisions the IPCC has invited criticism from those quarters where no consensus exists.

Framing climate change more broadly would open up the possibility of political action in more location-specific areas in terms of, among other things, regional adaptation with respect to improving air quality and energy services for the disadvantaged (Beck et al., 2014). Because the variable impacts of climate change are location-specific, it means different things to different countries, and to the different socioeconomic groups within those countries. Nonetheless, the fact remains that climate change is a global problem which, sooner or later, touches 'everything and everyone everywhere' (Moser, 2016). Because of this, it is one of those contentious issues that attracts all kinds of unfounded claims and frames. Framing climate change and its manifestations, sea level rise for example, has become a popular pastime for many of the actors and institutions that appear in this story.

Sea level rise is commonly associated with the implications of global warming, giving it a contentious edge. This is not unlike the associations made between nuclear power and the proliferation of nuclear weapons, and Muslims and terrorism. For example, in their book *Framing Muslims* (2011), Amina Yaqin and Peter Morey described the gap (post-9/11) that exists between the representation of Muslims – as 'homogenous and zombie-like' and ready to be whipped into a frenzy at the slightest disturbance to their supposed backward self-worldview – and reality. A biased worldview is established by emphasising particular aspects (the biased observation that all Muslims are terrorists because of the experience of the terrorist attacks on the United States on 11 September 2001, for instance) while at the same time omitting or deemphasising others. The extremism and violence of factional splinter groups of politicised Islam were amplified and projected onto Muslim communities *en masse*. Likewise, a planned or managed retreat from the coastline to protect against sea level rise and storm surge might be framed in the broader context of climate change, which remains a profoundly contentious issue, while its foremost utility resides in a common-sense rationale to protect citizens, housing, and infrastructure from extreme climate events.

The term 'climate change' is a politico-media equivalence frame in itself. It is a euphemism for global warming. The most illuminating example of relational theory, relevant to this story, was demonstrated by American political strategist Frank Luntz (2002) in a recommendation to the Republican Party in the Fall of 2002. Luntz advised that the term climate change was less 'frightening' than global warming. He maintained that the publics perceived global warming to be more serious than climate change; as such, there would be greater acceptance of inaction on environmental policy if framed under the banner of climate change. The mental representation of a change in the weather was perceived to be more palatable than the image of an overheating planet. The term 'climate change' was (and supposedly is) more ambiguous and less frightening than 'global warming'.

Understating the seriousness of global warming by framing it as climate change would presumably allow Republicans to shuffle climate change policy investments to the bottom of a list of public policy priorities leaving them free to pursue their professed neoliberalism and 'business as usual' ideologies. We can acknowledge then, that the term global warming has been quite successfully framed as climate change. This creates some uncertainty as to what climate change actually means. Whereas global warming is vivid in its description of an overheating planet, climate change to the uninitiated might simply mean a relatively innocuous change in the weather.

Framing climate change might also extend to framing the parameters for its prevention. It is worth pointing to the greenhouse gas (GHG) emissions reduction targets pledged at the COP21 in Paris, France, in November/December 2015.[2] Pledges to reduce emissions are referred to as Intended Nationally Determined Contributions (INDCs). Several parties to the conference nominated 2005 as their baseline year. For example, the United States: 26–28 per cent GHG emissions reduction between 2005 and 2025; Australia: 26–28 per cent GHG emissions reduction between 2005 and 2030; Canada: 30 per cent GHG emissions reduction between 2005 and 2030; and, New Zealand: 30 per cent GHG emissions reduction between 2005 and 2030.[3] Of significance is that the largest total anthropogenic GHG emissions in human history were recorded from 2000 to 2010 (IPCC, 2014a). Emissions were then temporarily constrained as a consequence of the global financial crisis (GFC) in 2007–2008. Choosing a year in the period in which emissions were the highest in human history, as a comparative baseline year, was a strategic decision to optimise the probability of achieving pledged emissions targets. But this could also be seen as a ploy to frame the perception that emissions reduction was much higher than it actually was, and thus overstate actual progress.

Because INDCs are discretionary targets (no enforceable global standard for emissions reduction currently exists), significant disparities ensue between the strategies adopted by different countries. For instance, rather than committing to an emissions reduction target in the first instance, China pledged that its CO_2 emissions (not GHG emissions which covers a broader spectrum of gases) would peak around 2030. It would then go about reducing CO_2 emissions per unit of GDP by 60–65 per cent from 2005 levels. Other offsets were applied such as increasing forest stock. Mindful that China is the world's foremost contributor to total global CO_2 emissions, which also makes it the world's foremost driver of climate change. Paradoxically, China is also initiating a concerted transition to cleaner energy sources.

Although nominating 2005 as the baseline year could be considered a sensible approach, it could also be construed as a crude politrick (discussed foregoing) as CO_2 emissions had already been reduced from their highs in the mid-2000s by a severe global economic downturn in 2007–2008. In short, the baseline years set for both the pre-industrial period and emissions targets are somewhat dubious in the sense that manipulating them can dramatically alter the actual amount of emissions reduced. Moreover, because different countries produce vastly different amounts of emissions, reducing emissions in percentage terms relative to divergent baseline years and target years makes little sense in terms of a genuine collective endeavour to remain within a global carbon budget. China's commitment to an INDC, which is in effect a non-commitment to directly reducing GHG emissions per se, is a case in point. More broadly, it is evident that throughout the process of framing INDCs, in the context of diverse political and economic drivers, there is the risk of losing sight of 'the objective' – that is to prevent dangerous climate change. Assuming, of course, that the scientific reality of dangerous climate change and its implications were empirically framed from the beginning of negotiations.

A study by Inga Schlichting (2013) found that since the early 1990s industry actors have employed three quite distinct frames to express their strategic political communications about matters relating to climate change. The scientific uncertainty frame was the dominant climate change communication in the early to mid-1990s. Its principal objective was to cast doubt on whether climate change was occurring and whether it was caused by CO_2 emissions. The rationale being that in the absence of scientific proof any action to mitigate CO_2 emissions was unjustified, irresponsible, and a waste of taxpayer funds. In terms of influencing public opinion, the emotive issue of wasting taxpayer funds was an exceptional card to play.

Post the publication of the IPCC's initial report, the *First Assessment Report* (FAR; IPCC, 1990), and subsequent IPCC reports, and as the fossil-fuel industry became embroiled in the climate change debate, from the mid-1990s to the early 2000s the frame shifted toward economic competitiveness and the negative socioeconomic impact climate change policies would have on economic growth. This particular frame remains the number one go-to politrick because it draws on the emotive issues (once again) of jobs and growth.

The current dominant frame – largely worldwide from the mid-2000s – is one of 'industrial leadership'. This frame adopts the old idiom, 'if you can't beat 'em join 'em'. It propagates the communication strategy of acknowledging that industry is also responsible for environmental protection and thus climate change; however, the solution is technology innovation, something the industry itself is best equipped to develop and provide. As observed with the propagation of media campaigns extolling the virtues of the energy sector's contribution to safer, cleaner, and more efficient energy supply.

I submit that a blend of all three frames is currently in play: uncertainty, economic competitiveness, and industrial leadership. They have been recycled and used selectively by industry and also by politico-elites (in different industries, contexts, and countries) as a matter of convenience and opportunism. The aim of which is to assimilate the mental models of the publics in the quest for public advocacy. Of course, from a politician's worldview, it could also be argued that the mainstay of such frames is the incumbent government's paramount priority to achieve 'year on year' economic growth. An argument politicians find difficult to oppose because of its association with employment and prosperity. The mainstay of industry is profitability which is also the mainstay of economic growth. The default mental model of the politico-elite, then, is such that economic growth is a product of industry profitability. Climate change policy is often framed as the archenemy of industry profitability, and the capitalist ideal.

Should all frames fail to convince the publics that climate change is either being diligently dealt with or climate change action should be deferred, politicians will revert to framing a distraction: draw the publics' attention away from the issue of focus by hyping up an alternative one. A popular diversionary issue of the late twentieth century and early twenty-first century is terrorism; or the fear thereof (Altheide, 2017). Politicians harness the power of mass media, and especially social media, to propagate the fear frame and expand the fear discourse as a prevailing means of social control. Slogans are packaged and presented, such as Donald Trump's 'keep America safe' which, according to David L. Altheide (2017), proved decisive in his victory in the 2016 U.S. presidential election. This is not to say that terrorism is not an important issue; however, the 'politics of fear' continues well after the fear has subsided. Perhaps ironically, one might also presume that climate change will become a primary motive for conflict, or terrorism as the case might be, as resource-poor countries and disenfranchised groups fight for control over scarce resources like water, for instance. These kinds of linkages are not easily verified, however (see for example Selby, Dahi, Fröhlich, & Hulme, 2017).

Framing scientific uncertainty

Climate change economist and academic, Richard S. J. Tol (2014), described climate change as 'a long-term, global, uncertain problem' (p. 124). No doubt the scientific uncertainty frame has left an indelible stamp on climate change per se, as noted by Geoffrey Heal and Antony Millner (2014). Independent philosopher of science Jerome R. Ravetz (2011) has argued that there are profound problems relating to the management of the inherent uncertainty in science and how this translates into policy. He maintained that the uncertainty attribute could not be resolved by more elaborate quantification: meaning that more extensive, sophisticated, or rigorous analyses will not dispel the element of uncertainty in the science. He pointed to the gap between science and policy; 'the languages, their conventions and their implications', as reasons why the translation of science into policy was problematic. The vernaculars and ideologies of science and policy (and moreover, of scientists and policymakers) are often incomprehensible to one another – and thus, most probably, to the publics as well.

Critics have observed that the exact wording in the *Summary for Policymakers* (the main focus for journalists, politicians, and policymakers) in IPCC reports is subject to negotiation and must be agreed upon by participating party countries. Political analyst and author, Andrew Heywood (2014), goes further to declare the *Summary for Policymakers* a politically negotiated document that occasionally omits contentious judgements found in the main body of the report.

Moreover, government representatives of oil-producing countries regularly endeavour to weaken statements in IPCC reports (that call for greater action on climate change) by emphasising the uncertainty in the science. A point to note is that participating party countries retain the power to veto certain aspects of IPCC reports based on the consensus principle.

Consequently, the climate change imperative might be framed in a way that is subordinate to certain geopolitical interests and sensitivities. By extension, it might be politicised to the extent of framing the legitimate risk of climate change as inferior to the scientific reality of the risk if the economic fallout is deemed to be unacceptable. Whether this purports to a false critique, political manipulation, or a case of unrealistic optimism, is a question not easily answered. Distinguishing the science from the politics is a major challenge and a longstanding one. The danger, however, is that a framing bias has the potential to place citizens in a false sense of security. For instance, the decision to build in a coastal zone vulnerable to sea level rise, coastal erosion, and storm surge inundation, based on a government approval to do so, or in a position of frustration in which government devolves itself of the responsibility and liability for approving such development. Mindful that development of this kind continues at a frenetic pace worldwide. Finally, is the science or the decision-making model the main source of uncertainty in decisions to implement climate change policy? Leonard Smith and Nicholas Stern (2011) addressed this question:

> One can rationally choose to act under uncertainty; indeed, [climate change] inaction is in this case a decision of substantial consequence. Even scientific sceptics will agree that action might be called for unless they can establish that the risk is, in fact, very small. Naturally, an anti-science lobby might still argue otherwise. A lack of certainty provides no rational argument against action.
>
> (p. 4820)

Conclusions

I have defined framing as to consciously or unconsciously sway an audience's preferences by understating, overstating, misstating, and/or omitting particular points of reference in a communication to create a preferred impression. Sophisticated politico-media framing effects have been implicated in influencing public opinion to the extent of 'manufacturing consent', especially when opinion lacks conviction and is therefore more easily swayed. The demands on communicators of the climate change imperative – primarily scientists and the politico-elite – are invariably high (Rayner, 2018). The stark reality of the climate change science, should the impacts of climate change prove to be 'high-end', should be communicated honestly to decision-making politico-elites and the publics. The challenge is, how to frame disturbing findings with a 'sense that a meaningful collective, societal response, and ultimately a dignified future of some kind, is possible' (Rayner, 2018, p. 379).

Further to this challenge are covert endeavours designed to scupper the kind of climate change communication described above. According to research by Robert J. Brulle (2014), conservative foundations in the United States, and those with a politico-economic interest – Koch Enterprises and ExxonMobil, for example – have poured hundreds of millions of dollars into the climate change counter-movement (CCCM). A social network composed of several thousand individuals with links to over 100 organisations dedicated to maintaining a frame that would confound the publics' understanding of the climate change issue, with the broader aim of institutionalising a delay to any meaningful climate change action (Brulle, 2014; Farrell, 2015). This less than admirable subterfuge obfuscates the kind of 'honest' communication required, drawing scientists into an unending debate. As Sven Ove Hansson (2018) observed, climate science should be communicated to the publics in a way that is independent of the denialist ruse, and not as a reflexive counterargument to it.

Notes

1 From *Star Trek* Season 3, Episode 7, *Day of the Dove* 1968: directed by Marvin J. Chomsky; written by Gene Roddenberry.
2 The 21st meeting of the Parties to the United Nations Framework Convention on Climate Change (UNFCCC).
3 At the time of writing 174 of the 197 member countries (Parties to the United Nations Framework Convention on Climate Change) had ratified the Paris Agreement.

3 Mental-modelling

In search of a theory

Sarah E. Fredericks (2014) pointed to the body of literature in the philosophy and sociology of science that supports the idea that facts and values are not completely separate entities: ‘Rather, one’s own ethical values; the values, assumptions, terminology, and methods of one’s epistemic community; and physical conditions shape one’s scientific and technical assessments of the world’ (pp. 6–7). This puzzle of how meaning is created and new concepts are formed, and how these new concepts relate to existing ones, has perplexed philosophers for centuries. Quite clearly, in the instance of creating meaning and forming an intellectually valid mental model the onus is on both the encoder and the decoder. As Johnson-Laird (1983) explained:

> All explanations must take something for granted. When one person successfully explains something to another, what is conveyed is a ‘blueprint’ for the construction of a working model. Obviously, a satisfactory blueprint for one individual may be grossly inadequate for another, since any set of instructions demands the knowledge and ability to understand them. In the physical sciences, for example, explanations are couched in the language of mathematics; and the average person has a tendency to avoid the language of mathematics.
>
> (p. 5)

In fairness to all, although the language of mathematics represents the purest form of explaining complex phenomenon such as climate change, deciphering it can be a challenge. One might argue that it is the publics’ responsibility, both individually and collectively, to assume the roles of citizen, politician, scientist, and saviour of the planet. In consideration of the complexity of the climate change challenge, the long-held idea of specialisation is anachronistic and counterproductive to the acquisition of knowledge. *The phenomenon itself determines the knowledge required to understand it.* Parts of it cannot be simply brushed aside because they do not bode well with one’s disciplinary division, interest, or bias. The aim for the layperson is such that he or she forms an accurate mental model of the object, scenario, or phenomenon – in this instance, climate change – which does require some understanding of its scientific, political, and sociological aspects. That said, have we become so transfixed on the need for specialisation, that it is no longer expected that the layperson comprehends a single field of it? As outlined at the beginning of this story, the problem of climate change is multidisciplinary and as such, the solution requires interdisciplinary enquiry. Jay W. Forrester (1971) described a mental model as a mental image of the world that you carry in your head. The downside is that it is composed of 'only selected concepts and relationships' used to represent reality. He further asserted that all laws passed and executive actions implemented are done so on the basis of a choice among alternative models. He also cautioned that mental models are fuzzy, incomplete and imprecisely stated. Similarly, Johnson-Laird (1983) maintained that there are 'no complete mental models for any empirical phenomena' (p. 4). He described mental models as only slightly more than high-grade simulations; useful to the extent that the picture perceived is accurate. How one determines this is unclear as no model-independent test of reality exists (Hawking & Mlodinow, 2010).

The idea of a mental image that purported to create meaning was perhaps first posited by Austrian philosopher Ludwig Wittgenstein. In his endeavour to define the nature of logical truth, Wittgenstein developed what is commonly referred to as Wittgenstein's picture theory of meaning. The picture theory of meaning was central to Wittgenstein's early philosophy. It was first articulated in his book, *Tractatus Logico-Philosophicus* (1922); parts of which he later contradicted (Richter, 2014). According to Wittgenstein's theory, the role of language is to paint pictures (a model of reality and the objects in the reality) of possible states of affairs – a perceived state of what the world is or could be. Wittgenstein declared that 'a picture is a model of the reality, and to the objects in the reality correspond the elements of the picture: the picture itself is a fact (in the sense of its logical form)' (Wittgenstein, [1922] 1999, p. 10). However, in the case of reality, the picture may or may not correspond with the fact, and be true or false accordingly. That is, a proposition (or sentence) about the world depicts reality by painting a state of affairs in a particular way according to the artist. And it is only by picturing reality that individuals acquire a sense of the meaning of the proposition. Note that if you find Wittgenstein's *Tractatus Logico-Philosophicus* austere and contradictory, you would not be alone.

The first documented reference to a mental image as a model per se, that prompts a mode of action, can be possibly accredited to Sir Frederic C. Bartlett, a Professor Emeritus of Experimental Psychology at Cambridge University in the 1930s. In his book, *Remembering: A Study in Experimental and Social Psychology* (1932), he referred to remembering as 'an imaginative reconstruction, or construction, built out of the relation of our attitude towards a whole active mass of organised past reactions or experiences, and to a little outstanding detail which commonly appears in image or in language form' (p. 213). Bartlett's primary focus was to formulate a theory of remembering, which he defined as a complex mental rendering of life (composed mainly of mental images) in that moment that helps determine our daily modes of conduct.

The first proponent of the modern mental model theory was British philosopher and psychologist Kenneth J. W. Craik. In his book *The Nature of Explanation* (1943), Craik made a remarkable analogy of the human as a processor of information, similar to that of the modern-day computer that had not been invented at the time of his writings (as noted by Johnson-Laird, 1983). Craik made noteworthy parallels to the predictive powers of computer modelling:

> By a model we thus mean any physical or chemical system which has similar relation-structure to that of the process it imitates. By 'relation-structure' I do not mean some obscure non-physical entity which attends the model, but the fact that it is a physical working model which works the same way as the process it parallels, in the aspects under consideration at any moment.
>
> (p. 51)

Craik also made reference to reacting to 'future situations before they arise' and thus a cognitive pre-emptive function based on past and present information – the inference of some form of predictive power. Predicting the future is the essence of human existence. In this sense, the mental model is based on probability theory. Similar, in effect, to a computer model used to predict extreme climate events. And in the same way, superior probabilistic mental models that exhibit the qualities of hindsight, insight, foresight, and imagination have not only made great thinkers and leaders, but have also made important contributions to the survival and extraordinary success of the human species.

A mental model defined

In simple terms, a mental model can be thought of as a predetermined infographic of how to think and act. The continuance of the human species might ultimately depend on mental-modelling aptitude. Mental models are future-oriented in the sense of the thinking and acting that they prescribe. Their purpose is to realise a goal (Kaku, 2014), which could relate to the past, present, or future, or all of these. For instance, the goal to better understand climate change today by recalling the past when the mean surface temperature of the Earth was 3–4°C warmer to simulate what the world might look like in 2100.

Perhaps the most influential and succinct mental model, to explain deductive reasoning, was that proposed by Johnson-Laird and several of his co-authors (Johnson-Laird, 1983, 2013; Johnson-Laird & Byrne, 1991; Johnson-Laird, Khemlani, & Goodwin, 2015). This model is based on the rationale that 'the conclusion must be true if the premises are true' (Johnson-Laird, 2013, p. 1). Johnson-Laird (2013) proposed three sequential stages of thinking when reasoning. The first stage is to comprehend the premises, the starting point of deduction, through the use of general knowledge and knowledge of language. Johnson-Laird acknowledged, however, that there has been no formal account of how modellers translate their perceptual observations derived from the premises into mental representations – a conundrum that has mystified philosophers for centuries.

The second stage of developing a mental model is to formulate a putative conclusion. His reasoning was that a prudent conclusion could be drawn by developing associations that were not already explicit in the premises; in essence, a composite of the premises that moved closer to a plausible conclusion. The final stage entailed a test of the validity of the putative conclusion. This required a search for alternative models of the premises – of other possibilities – in which the putative conclusion was false. If such a model is found then the putative conclusion is false, and the search for another conclusion and the procedure for its validation begins. A plausible conclusion, then, is one where there are no other models that might falsify it, based on the rationale that it must be true in all cases that the premises are true. Of course, an implausible conclusion can occur if the hypothesis is mistaken for a true premise,[1] or used as a true premise, to counter an inconvenient truth. Alternatively, the premises might be too complex for the modeller to arrive at a plausible conclusion. Climate change, for example, is considered a 'wicked problem' (see for example Curran, 2009; Levin, Cashore, Bernstein, & Auld, 2012) that is scientifically complex and difficult for the layperson (or non-scientist) to grasp (Boschetti, 2012; Chen, 2011). In which case the modeller might simply be left perplexed.

Johnson-Laird concluded that the most formidable obstacle to achieving a plausible conclusion is that humans are cognitive misers, or deductive satisficers. As soon as they find a model that is compatible with the information in the premises they are satisfied and the mental-modelling process is complete. There is a disposition to avoid any further energy expenditure. Individuals might by chance validate a seemingly compatible model in the first instance; however, by doing so they fail to explore all the alternative possibilities that might comply with the premises. Thus, they draw conclusions that only hold for some of the premises. The consequence of which might lead to a deductive failure and a biased conclusion. Such deductive failures, the failure to acknowledge the probability of a shift in the Earth's climate, for instance, could prove 'catastrophic'.

German sociologist Ulrich Beck (2014) posed the following thesis: 'Climate change is the embodiment of the mistakes of a whole epoch of industrial capitalism, and climate risks pursue their acknowledgement and correction with all the violence of the possibility of annihilation' (p. 169). Beck then tempered the use of the word 'annihilation' with a lucid reasoning: 'The apocalyptic climate change is, maybe, the best documented fact we have ever had' (p. 173). He goes on to declare that this knowledge has not translated into action but a denial of the facts. These are strong words. Beck has framed climate change as a genuine threat to the survival of the human species and is clearly dumbfounded by the insipid response. Plainly, the science that informs the premises that informs the mental models formed has been lost in translation, somewhere between discovery and denial.

Filling the knowledge gap

It is reasonable to surmise that individuals hold a range of imperfect to near perfect mental models of the climate change problem. These are based predominantly on their physical experience of its manifestations and their respective knowledge of the science. In light of the often-indifferent response to climate change one could also surmise that the publics' knowledge of the science is inadequate. As Saul, Sherwood, McAdam, Stephens, and Slezak (2012) put it: even in developed countries, despite a vibrant media, high levels of education, and widespread access to information via the Internet, the gap between public opinion and scientific reality is both 'stark and disturbing'.

As stark and disturbing as this gap might be, grasping the scientific reality of the climate change problem might not be so easy. Fabio Boschetti (2012), an applied mathematician, argued that because of the many interacting variables, humans are limited by their cognitive capacity to accurately assess the long-term dynamics of climate change. Xiang Chen (2011), a cognitive psychologist and philosopher of science, argued along related lines. He pointed to an error in people's ontological assumptions – a person's subjective assessment of the nature of reality – that might explain the widespread misunderstanding of climate change. He proposed that humans have a tendency to treat climate change as a static object rather than a complex dynamic process, thereby misinterpreting its cumulative effects and growing consequences. The fact that policy-makers continue to set gradual greenhouse gas emissions reduction targets that range in percentage terms (currently between 25 per cent and 50 per cent for advanced liberal democracies) and over several decades (generally over a period of 25 years) is evidence of this error. Greenhouse gas concentrations will continue to accumulate irrespective of incremental reductions on this scale.

What is problematic about the abstract climate change object is that humans understand abstract concepts through the bodily experiences associated with them – referred to as 'embodied cognition theory'. Embodied cognition theory posits that 'the mind cannot work separately from the physical world; that senses provide the bridge to both our unconscious and our conscious thought process' (Lobel, 2014, p. 11). Most humans, however, would not have experienced climate change, or recognised that they experience climate change, through a bodily experience, even though it can be empirically observed, measured, and quantified.

This aligns with a 'matching bias'. Defined as the propensity to interpret as relevant information that matches the lexical content in a statement or what is observed, and reject logically relevant information that fails to match that interpretation (Evans, 1998). Matching bias is most commonly associated with abstract content and is understood as an intuitive heuristic, meaning that it involves intuition rather than reflective reasoning. Reflective reasoning is associated with mental model validation. Note, however, that mental model validation might be viewed as counterintuitive if it requires an action based on a theoretical threat, such as climate change, that cannot be measured in terms of a human lifespan.

Climate change is often only visible at the margins. For example, the low-lying small island nations of the South Pacific and the Polar Regions where sea level rise and ice melt can be observed first hand. Otherwise climate change is flawless in form and invisible to the eye. One is reminded of an Oscar Wilde quip: 'Before Turner, there was no fog in London'. Although science has painted a clearer picture of the implications of climate change, it remains veiled in abstractness and mired in politicking. From the lounge chairs of suburbia climate change 'involves invisible gases moving through invisible air to trap invisible heat, with computer models telling us that this will cause a planetary catastrophe' (Nicholson & Wapner, 2015, p. 4). Climate change, then, is not only a 'wicked problem' but also an 'invisible' one. It is encoded in the language of science, framed in the language of politics, and evaluated in the language of economics. The irony of it all is that severe and abrupt climatic change has not been experienced in living memory, but could prove fatal because of this.

Caroline Gauffroy and Pierre Barrouillet (2014) hypothesised, based on Johnson-Laird's (1983) mental model theory, that because mental models are transient cognitive representations held in a limited-capacity working memory,[2] 'cognitive economy' (the cognitive miser as first described by Taylor, 1981) would lead to an implicit format of the alternatives – an implied list of all the possibilities that an individual could imagine. The burden of a limited-capacity working memory constrains the complexity of the mental models that individuals can construct. However, if one was to engage a more time-consuming and effortful cognition, the initial implicit set of mental models could be improved upon, resulting in a more explicit set to choose from. The heuristic system is automatic, unconscious, and pragmatically cued to a default mental model that comes quickly to mind. Analytic reasoning, however, is characterised by a more conscious and effortful endeavour to process the underlying semantics of a proposition by constructing a more explicit mental model.

Robert Northrop and Anne Connor (2013) affirmed that human cognitive capacity has not evolved to the extent that it can effectively cope with complexity and complex systems. Having descended from short-lived hunter–gatherers whose principal threat was from sudden attacks from apex predators, other human adversaries, and disease and starvation, humans were not equipped for modern-day, comparatively slow-moving threats such as overpopulation and climate change. As a consequence, when something harmful to humankind occurs sufficiently slowly – relative to the scale of human life – it fails to trigger a 'meaningful, organised, social response' as a precautionary measure until some form of martyrdom transpires. This is referred to as the Social Action Rate Sensitivity Law (SARSL). For example, the bolstering of seawalls, dams, and dikes typically takes place in the aftermath of their failure, not in anticipation of it, and perhaps due to loss of life. Likewise, a planned retreat from the coastline might not be implemented until the entire coastal zone is inundated – perhaps for the second or third time.

Conclusions

Evidence suggests that how an issue or scenario is framed, in this case climate change, is instrumental to the mental model formed. Subsequently, the mental models that individuals hold might be accurate, reasonably accurate, or totally inaccurate. Moreover, the downside of ruling incumbents holding to totally inaccurate mental models is comprehensive.

People tend to process information in a way that is consistent with their pre-existing beliefs and values. A person's default mental model of how they perceive the world influences how they understand the world. Accordingly, heuristic processes produce a default mental model that is easily brought to mind. That is, people commonly reason on familiar associations: automatically linking objects, scenarios, or phenomena together. In light of this, most people are unfamiliar with the scientific phenomena of climate change, in particular, recognising its physical attributes as distinct from what they would normally associate with natural fluctuations in the weather. This suggests that more effort is required to bring a plausible model to mind. More broadly, with this in mind, one might surmise that the default mental model held by most of us is unlikely to yield a truthful representation of how the world actually works.

The climate science that informs the climate change mental model is explicit but not complete. Inklings of uncertainty persist. The default mental model of waking to an idyllic sunrise over the ocean diminishes the compulsion to validate a precautionary model, such as a planned retreat from the coastline – a natural buffer that has intrinsic amenity – that would protect against intensifying storminess, sea level rise, and storm surge. Current thinking mainly views extreme climate events as rare and the damage a cue to unite and rebuild. There is the 'belief' that hurricanes, like Harvey and Irma, that caused unprecedented flooding and destruction in coastal zones throughout the Caribbean and United States in August/September 2017, happen only once in a lifetime if at all. This is exemplified by a 'willingness to pay' a high price for a house by the sea. Assuming, then, that the default mental model of a house by the sea appeals to the 'majority' – the 'tyranny of the majority' so to speak – the politically imposed precautionary adaptation policy of a planned retreat, implemented in response to predictions of intensifying storminess and sea level rise, necessarily conflicts with other policy goals, such as upholding the democratic principle, property rights, and the freedom to choose.

A recurring theme is that mental models are predictive mechanisms. They are future-oriented in terms of the thinking and acting they prescribe. But they are also usually incomplete, sometimes hazy, and have gaps that are often filled with theory and superstition or fiction. It is evident that people find it difficult to accurately mental-model climate change and often hold misconceptions about its causes and solutions that might lead to miscalculations of its implications. This is compounded by the uncertainty in the science that has, in the past, been framed to justify inaction. Positive correlations have been drawn between increased knowledge of the consequences of climate change and the consequent level of concern. It is reasonable to conclude, therefore, if an individual's mental model of climate change is deficient – knowledge gaps exist and/or it is susceptible to framing effects that could alter what is known to be true – then his or her corresponding thinking and acting will be similarly deficient. Likewise, a deficient mental model of climate change can only lead to a deficient climate change policy.

Notes

1 For this story, a 'true premise,' apart from a given like the sky is blue, for example, is something not taken for granted in the first instance, irrespective of how or by whom it is presented. Identifying a true premise requires the actor to: (1) ascertain that it derives from a reliable, or even empirical, source; and (2) recognise that it is supported by sound reasoning even if it appears intuitively probable or improbable.

2 Working memory differs from short-term store (STS) in that it has the capacity to transform the information held: whereas STS involves only the short-term storage and reproduction of information that is lost if not worked on. For example, working memory is engaged when a user is required to memorise parts of information or information in a different order that is held in STS. Evidence suggests that working memory and STS are separate but interconnected constructs (Schneider, 2015). Moreover, in accord with Baddeley and Hitch's 'working memory model' (Baddeley, 2000; Baddeley & Hitch, 1974), working memory refers to 'a limited capacity system' – with control limits on both storage and processing capacities – that allows for the temporary storage and manipulation of information used in complex tasks such as comprehension, learning, and reasoning.

4 The General Model of the Policymaking Process

Policy is the fruition of a competition among ideas, ideologies, and interests that impel, but also impede, a political system. Policy paralysis can potentially shut down government as the United States Congress has demonstrated in recent years. Policy as it is traditionally applied in political science can be defined as the administrative rules and principles that bureaucrats use to organise and structure the execution of their administrative powers (Mann, 2013). It is the instrument of governance seen as an authoritative response to a public issue or problem that directs public resources in a preferred direction.

Policy is made in response to decisions of government. Governments have theories prescriptive of the mental models they hold about how the world ought to be and about how humans should think and act. The purpose of policy is therefore to encourage thinking and acting that aligns with such theories and discourages thinking and acting that does not. Theoretical assumptions about reinforcing or modifying human thinking and acting cannot be effectively tested pre-implementation; hence, policy incorporates probabilistic predictions about future commitment and compliance. Subsequently, although policymaking is prospective, the efficacy of policy only becomes clear in retrospect.

Public policy, according to *Black's Law Dictionary*, is broadly defined as 'principles and standards regarded by the legislature or by the courts as being of fundamental concern to the state and the whole of society'; and more narrowly, as 'the principle that a person should not be allowed to do anything that would tend to injure the public at large' (Garner & Black, 2014, p. 1426). Public policy, then, with respect to individuals and society as a whole, infers a universal 'duty of care' both in a legal sense and as a matter of principle, in terms of norms and values and a moral obligation to safeguard the rights and wellbeing of the publics in general. Thomas A. Birkland (2015) extended this to define public policy as: 'A statement by government of what it intends to do such as a law, regulation, ruling, decision, order, or a combination of these. The lack of such statements may also be an implicit statement of policy' (p. 9). For instance, climate change policy inaction might convey the impression that government does not consider climate change a genuine threat.

Public policy is thus enforced through a regulation, ruling, or law; or more succinctly, statutes (legislation). Statutes encompass 'the body of law that deals with the powers, rights, and obligations of the government and the governed' (Butt, 2004). For instance, a climate change adaptation policy, such as a planned retreat, might be enforced through environmental law, such as the Precautionary Principle, or through tort law. Under the former, government would enact a law to either prohibit development in, or withdraw assets from, vulnerable coastal zones. Under the latter, interested or aggrieved parties would seek a judgement and/or remedy by testing their rights in a court of law.

The policymaking process (or policy cycle) was defined by James E. Anderson (2014) as 'a sequence of functional activities beginning with problem identification and agenda formation, and continuing through evaluation of policy, which might result in the continuation, modification, or termination of a policy' (p. ix). Suffice to say, policymaking involves the coordination and balance of a multiplicity of interests. Policy practitioners and other interested parties are charged with bargaining and balancing these interests to decide on a policy choice. The aim is to raise enough support for a policy decision and thus ultimately, policy design and implementation.

Politics is central to policymaking. Politics can be defined as processes and actions – enabled by the relations between people, groups, and resources – necessary for a political system to function. Alternatively, Heywood (2014) defined politics as power: the capacity to realise a preferred outcome 'through whatever means'. He elaborated by describing power as the capacity to modify the behaviour of others in a 'manner not of their choosing'. Plainly, to purposefully prescribe certain mental models of how citizens should think and act. Citizens are incentivised by rewarding compliance, and disincentivised by punishing non-compliance. The reward for compliance is not being punished for non-compliance.

Democracy delimits the extent of the State's power. Citizens are afforded certain rights, such as property rights, that protect against the excesses of government. Democracy and the rule of law rest on this. In this way the limits to government power are self-imposed and thus, must be self-enforced (Weingast, 1997). Knowledge is a precursor to the government act of power through policy; therefore, the correct use of knowledge must be self-imposed and self-enforced. As Ross Garnaut (2011) explained: if the publics were erroneously persuaded that the climate science was misleading then they are denied the knowledge to exercise sound judgement which is their democratic right.

There are liberal democracies that are guilty of disseminating misinformation, not correcting misinformation, and endeavours to deny the publics of climate change knowledge. This fuels an ill-founded debate about the authenticity of the climate change problem. Reports in news media of periodic endeavours to discredit the science and prohibit the use of references to climate change in policy documents and government websites, as well as science lessons in schools, are numerous in advanced liberal democracies (see for example Bourke, 2017; McGrath, 2017; McKibben, 2017; Rosen, 2017). Such endeavours commonly correspond to the political ideology of the communicator, and the government in office at the time. In an Australian context, consider the following excerpt that highlighted the precarious nature of the translation and transmission of scientific knowledge through a network of disparate and/or deficient mental models of the climate change problem:

> The organisational wing of the [Queensland] Liberal National Party [the Liberal National Party (LNP) is a coalition partner of the centre-right federal Liberal Coalition] has voted to 'remove environmental propaganda' on climate change from Queensland schools. A delegate at the LNP's three-day annual conference cited concerns 'false prophets' were poisoning children's minds. Delegates on the first day of the conference in Brisbane agreed to call on Education Minister John-Paul Langbroek to ban 'post-normal science' from the curriculum and exam materials. LNP Noosa delegate Richard Pearson, who moved the proposal, said climate change information was being used as a strategy in schools to promote a political view, and the science was not set. 'It's aimed at the heart of those false prophets who would poison the hearts and minds of our children in the Queensland education system,' he said.
>
> (Barrett, 2012, paras. 1–5)

This kind of thinking essentially exploits the methodology of 'post-normal science' as proposed by Silvio Funtowicz and Jerome R. Ravetz (1993) – post-normal science is appropriate when the attributes of systems uncertainties and decision stakes are high – to argue 'the science was not set'. Scientists have confirmed that warming of the Earth's climate is unequivocal. Moreover, as argued by Thomas C. Walker (2010), the paradigm mentality posited by Thomas S. Kuhn which purports to 'normal science', conflicts with the idea of an open society of scholars using an array of methods and theories in the quest for knowledge, as espoused by Karl R. Popper (1963) and his idea of 'falsifiability', and thus impedes the growth of knowledge. More succinctly, as Walker explained, Kuhn's model (and also Imre Lakatos's model of the progressive research programme) is fundamentally conservative and acts to suppress criticism and is therefore authoritarian. From a broader perspective, although Kuhn's ([1962] 2012) idea of 'normal science', as opposed to post-normal science, advocated a more structured methodology, his description of 'normal science' as 'dedicated to solving puzzles and truth seeking' appears indubitably universal.

A historical counterpart to the excerpt above (Barrett, 2012) is perhaps the Edict of 1616, a public policy presumably in the public interest but more likely in the interest of the Catholic Church at the time. It is a classic example of the suppression and fragility of scientific knowledge. This edict comprised an order by the Church to prevent Galileo from holding, defending, or teaching his theses (Copernicus's theory) that the Earth and the other planets orbit the sun, as opposed to the sun and the planets orbiting the Earth, a strongly held proclamation by the Church. Galileo's book, *The Starry Messenger* published in 1610, attracted public accusations of heresy that led to his appearance before the Roman Inquisition. Galileo went voluntarily in an endeavour to clear his name.

The Church censured two of his theories concerning the motion of the Earth, giving rise to the Edict of 1616. In essence, Galileo was ordered to reconcile his theories of the Earth's motion with that of the Bible's. Not dissimilar to that of the Liberal National Party's endeavour to denounce climate change science in accord with its particular ideological settings. Galileo stood against the hegemony of the Catholic Church and the imbalance of power at the time so that science could stand up to the forces that resist scientific knowledge and decide on matters of fact on an empirical and experimental basis.

Although times have changed since Galileo's stoush with theocracy, the prevalence of deficient mental models and the oppression and fragility of scientific knowledge has not – it matters not whether the agent of authority is a Theocracy or Bureaucracy. In a liberal democracy, the political system that underpins the functional aspects of society draws on the 'doctrine of liberalism'. This doctrine was founded on justice, fairness, and individual rights (Sandel, 1998). Such a democracy aspires to a political system in which democratically founded institutions are constrained by liberal-inspired constitutional arrangements, political practices and popular expectations that limit intervention from the government sector.

Suffice to say, the suppression of scientific knowledge could hardly be in accord with the principles of a liberal democracy; or to borrow from political philosopher John Rawls (2011), a 'well-ordered democracy'. Rawls (2011) described a 'well-ordered democracy' as a 'deliberative democracy', with the idea of citizens deliberating freely and exchanging views about political issues and questions of public concern being a fundamental countenance of democracy. He also emphasised that the role of governments was to inform and the responsibility of citizens was to be informed:

> Deliberative democracy also recognizes [*sic*] that without widespread education in the basic aspects of constitutional democratic government for all citizens and without a public informed about pressing problems, crucial political and social decisions simply cannot be made. Even should farsighted political leaders wish to make sound changes and reforms, they cannot convince a misinformed and cynical public to accept and follow them.
>
> (p. 449)

Being informed, of course, is not simply a privilege bestowed upon one but requires some endeavour. In an age when public cynicism is at a historical high, the motivation to make such an endeavour is perhaps wanting. Cynicism, and also knowledge of an issue or problem, is reflected in the publics' opinion. Public opinion is central to democratic theory because it lies at the heart of the people–government relationship. In communication theory, public opinion is fundamental to studies that examine the influence of mass media. In economic theory, public opinion is considered an important factor in influencing a country's development. It is pervasive in discourse about giving voice to the publics' will in the policymaking process. Equally so, public opinion can reduce policy to popularism as governments vie for the majority vote. A 'purely cosmetic policy', reduced to popularism or captured by vested interests, risks being drained of its efficacy and thus, also renders the supporting science immaterial.

As Garnaut (2011) lamented, because climate change policy impinges on strong interests it is rudimentary for vested interests to 'capture policy' and the evidential need for the policy to be overlooked. To capture policy refers to 'regulatory capture' and/or 'cognitive capture'. As defined by Nobel Prize winning economist Joseph Stiglitz (2013), the former refers to those regulators that regulate an industry from which they come and might return to, thereby treating that industry favourably. The latter refers to those regulators that simply need or desire to identify with a particular group. The mind of the regulator and thus the policymaking process is captured by those they regulate.

The Model

The General Model of the Policymaking Process (Figure 4.1), derived herein for the purpose of explanation, comprises a number of broad theoretical ideas drawn from the processes, actions, and actors involved in the policymaking process. The foremost of these relate to the translation and transmission of scientific knowledge, in the form of words, pictures, and probabilities, across the scientific, public, and political landscapes. The Model depicts scientific knowledge originating in the form of a report by the IPCC – the recognised expert mental model in this story. This knowledge, which might also be described as scientific reality, passes through a sequence of processes and actions – of encoding, decoding, and negotiating reference frames – and theoretically at least, it is ultimately incorporated into climate change policy design and implementation.

I apply the Model to the *Summary for Policymakers*; the main transmission of interest in IPCC Assessment Reports for journalists and policymakers (see for example IPCC, 2014c, pp. 1–32). The *Summary* comprises policy recommendations made by the IPCC. It is based on a comprehensive report that underpins the scientific and technical data as it explicates the current state of the climate change phenomenon and endeavours to predict outcomes based on an array of future scenarios. The efficacy of knowledge translation and transmission, however, is ultimately subordinate to a unique decision-making/debating ritual that typifies the parliamentary process. Arguably, this ritual lacks empirical foundation. Fought in the sophist tradition, a persuasive argument, contracted to the policy of party-preference, is mounted on either side. This might appear 'opportunistic or cynical' as suggested by Michael J. White (2012). And observed by Timothy Jenkins (2013): 'Sincere speech is the aim of scientific activity – the creation of a disinterested speaker, in contrast to the sophistry of a rhetorician – when outward signs correspond to a purified inner state' (p. 73).

I suspect that 'a purified inner state' is not so easily attained, as scientific man is prone to the same weaknesses as political man and economic man, and for that matter, politico-media man. No man is an angel as we shall later discuss. Nonetheless, it is presumed that scientific knowledge is presented in a neutral frame,

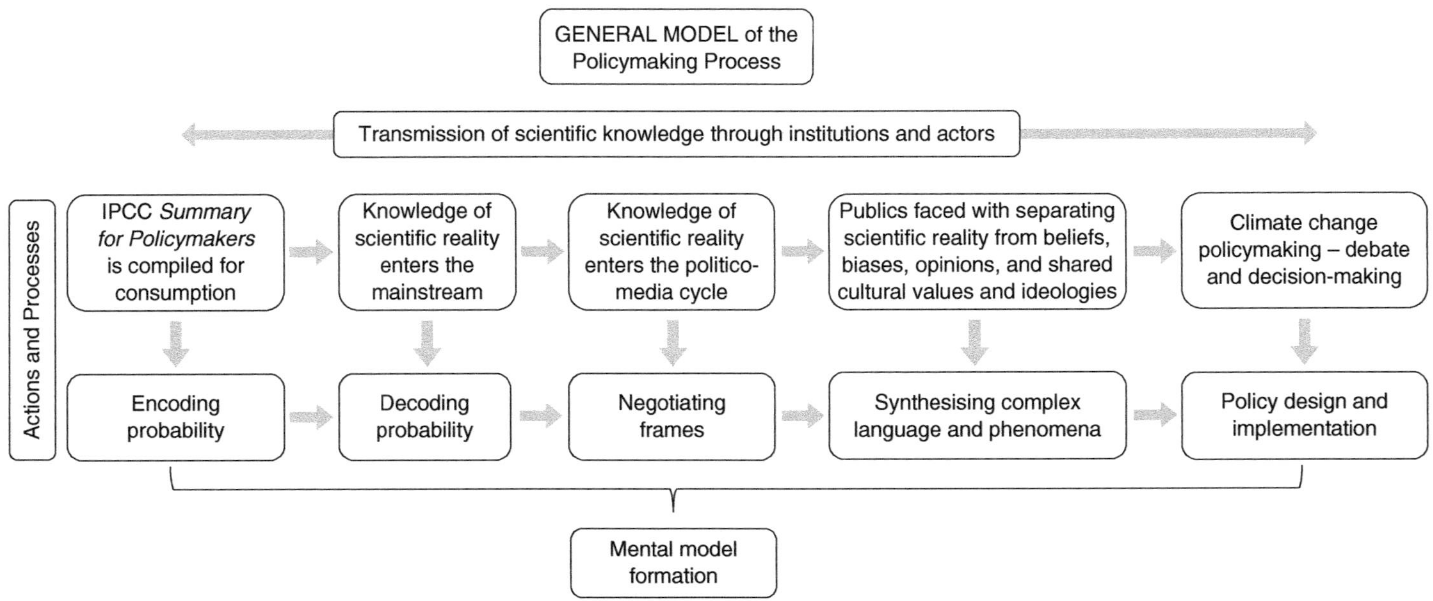

Figure 4.1 The General Model of the Policymaking Process.

even though it does not occupy a neutral space. That is, one must persuade oneself of the science through (1) understanding the science and (2) accepting the science as true belief. A particular inflection point (upper-centre in the Model) for either purification or opportunism and political cynicism, as observed by Jenkins and White foregoing, is when knowledge of scientific reality enters the politico-media cycle. It is now subject to both praise and persecution contingent on how closely it aligns with the ideals of party-preferred policy and other vested interests such as that of the publics' and those whose way of life and/or livelihood and beliefs it might disrupt.

Prediction is about the probability of matching outcomes with expectations. Decoding the probability expressed in degrees of certainty – the probability or certainty of a particular event occurring – as depicted in IPCC Assessment Reports is of particular interest and open to lengthy debate. Indeed, disparate mental models of the climate change problem, charged with the euphoria of political debate, might overwhelm the science and also the policymaking process to the point of policy paralysis – the nemesis of all political systems. In a similar way, acknowledged is the preponderance of scientific data that might impose an exactitude that further stifles such debate. Added to this is the publics' capacity to synthesise complex language and scientific phenomena and subsequently form an accurate mental model of the issue of focus, and thus provide the advocacy that contributes to policy success. There is also the probability of the party-preferred policy realising its predicted outcome. The outcome might fall short of expectations or exceed expectations or indeed, create other unexpected expectations.

Since probability theory is central to IPCC communications (Adler & Hadorn, 2014; Budescu et al., 2009; Budescu, Por, & Broomell, 2012), it follows that its comprehension is requisite to a commensurate policy response. However, several authors (see for example Hertwig & Gigerenzer, 1999; Stanovich, 2009; Tversky & Kahneman, 1983) have argued that the layperson might lack an understanding of the rules of scientific thinking and probabilistic thinking which accounts for much irrationality. Stanovich (2009), for instance, has claimed that the lack of knowledge of probability theory is a 'mindware gap' that is widespread. Extensive research has shown that our 'natural thinking tendencies' do not generate the correct estimates when processing information of a probabilistic nature. The human mind is not designed to operate according to the rules of probability and often violates the conjunction rule of probability theory. Furthermore, Nassim Taleb, in his popular book, *Fooled by Randomness* (2004), argued that it is the failure to grasp probability theory that leads people to believe that outcomes are non-random – humans strive to identify patterns where none exist. Subsequently, some degree of science literacy – public understanding of science – and knowledge of probability theory is indubitably helpful in processing climate change information.

It is apt to note that contrary to the foregoing and despite a general consensus, the concepts of heuristics and biases and their effect on probabilistic thinking have not gone unchallenged. For instance, Fintan Costello and Paul Watts (2014) pointed out that since the capacity to reason about probabilities is central to survival – and humankind has indubitably flourished – it follows that humans must do it 'extremely well'. Their research revealed that individuals do make probability judgements in accord with probability theory, and biased judgements are a result of 'random variation or noise in the reasoning process' (p. 463). In essence, the heuristics and biases described by Tversky and Kahneman (1974, 1983) are simply noise in one's recall.

Costello and Watts maintained that when this noise was removed, individuals' judgements do indeed concur with the requirements of probability theory. I would add however, that everyday complexity that requires probabilistic thinking is comparatively modest when weighed against the complexity of scientific phenomena like climate change. That said, the question of whether these heuristics and biases extend to scientific reasoning is also one of debate (French, 2016). The scientific method, for instance, could be viewed as a heuristic of sorts, albeit a premeditated one. As declared by C. Wright Mills (1959), who regarded empiricists bound by the constraints of a dominant theory or method, as 'methodologically inhibited'.

A language of images

Returning to the Model. Once scientific knowledge enters the mainstream it is framed for public consumption by various actors and institutions and thus exposed to the weathervane of public opinion. Navigating the path from prediction (in terms of probability) to policy (design and implementation) requires various actors to negotiate the frames of reference and communicate through various mediums. Discourse, for instance, is grounded in both words and mental imagery: 'The most powerful, meaningful, and culturally important messages are those that combine words and pictures equally and respectfully' (Lester, 2013, p. xi). For example, Kaku (2014) in his book *The Future of the Mind*, uses films as reference frames, such as *The Matrix*, *The Manchurian Candidate*, and *The Bourne Identity*, to more vividly and cogently create and convey his thoughts. The films themselves could even be viewed as the cinematic renderings of a reel of prefabricated mental models assembled in a prefabricated world. Moreover, contemporary society has embraced digitally enhanced imagery and Internet technology that is quickly evolving a language dominated by images – reinforcing the age-old idiom, 'a picture is worth a thousand words'. I suspect that words cannot always do justice to the image they endeavour to describe. Whereas an image speaks all languages which weds well with the universality of the Internet.

As one proceeds through the General Model of the Policymaking Process, graphical images come to mind. Graphical representations are central to presenting logical arguments and conveying scientific knowledge (Roth & Bowen, 2003). Perhaps more than previous efforts to convey a phenomenon, visual imagery has been used to communicate climate change. Birgit Schneider (2012) observed that the results of climate change research (such as the IPCC

Assessment Reports referenced in Figure 4.1) could not be communicated without the extensive use of visualisation: images of a changing climate in the form of maps, graphs, and futuristic simulations of ice melt are fundamental to framing climate change in endeavours to persuade the publics and the policymakers of the mounting threat. Indeed, effective imagery plays a central role in judgement and decision-making with regard to global warming by influencing how the stimulus is interpreted, the level of attention paid, and the action taken (Smith & Leiserowitz, 2012).

A study by Wolfgang Schnotz and Inga Wagner (2018) focussed on the conjoint processing of text and pictures in the context of performing specific tasks. The study involved a cohort of 204 students from grades five to eight in the German school system. Schnotz and Wagner found that because text provides more conceptual guidance for systematic comprehension than do pictures, initial mental model construction was more text-driven than picture-driven. However, when a specific task needed to be performed post the initial comprehension phase, students were more likely to call on pictures. The reason for this was because pictures allow fast and flexible access to task-relevant information, and can also be used as external cognitive tools on demand. This suggests that text provides a conceptual guide for comprehension in the first instance of mental model construction. In order to solve a specific task, the mental model can then be adjusted to the demands of the task through a detailed study of corresponding pictures. Schnotz and Wagner referred to this as 'adaptive mental model elaboration'. In short, pictures (or images) were fundamental to enhancing understanding and executing a task specific to its demands.

Images require information in order to evolve. Michael R. W. Dawson (2013) explained that visual perception becomes especially difficult when the goal is to create a realistic internal model (mental model) of the world. This is because visual representations are not information-rich enough to explicitly reconstruct the world's unique three-dimensional qualities. One solution to this problem, called the 'New Look', is to regard 'seeing' as a form of problem-solving. The 'New Look' approach is based on the premise that beliefs, expectations, desires, and general knowledge, all contribute to the visual experience by providing the missing information to the raw image so that individuals can complete a coherent model of their world. The idea being that with the imagery comes certain pieces of descriptive data, or visual knowledge essential to completing a three-dimensional mental model.

There are two fundamental types of imagery: internal or 'mental imagery' (mental pictures) and spatial or 'visual imagery' (ordinary pictures). A distinction between the two is the way they are organised by the brain. The brain is organised topographically, and as such, specific areas are activated in accord with specific cognitive functions (Kaku, 2014). Both visual imagery and mental imagery are thus essentially internal representations (also referred to as mental representations or social representations). The former derives externally from visual perception (seeing) whereas the latter is derived internally from introspection. However, the two have overlapping neural areas within the brain (Reisberg,

2013). Moreover, in the case of philosophy, the doctrine of representationalism holds that the mind comprises a system of mental representations about the world that portend to intentional content: whatever they are about, one's mental states concern intentionality (Neander, 2017). Note that in this instance, intentional mental states encompass all representational mental states: 'perceptions, beliefs, desires, memories, decisions, judgements, imaginings, ideas, emotions, reasoning and possibly sensations along with any representational attributes of emotions and feelings' (Neander, 2017, p. 145). One might thus argue that these qualities are directed toward a particular state of affairs, an object, scenario, or phenomenon. One might also view this system of mental representations as a language of images supplemented with the occasional subtitle only to enhance perception.

Mental imagery is manifestly a non-verbal mode of thinking. Psychologist and neuroscientist, Stephen M. Kosslyn (1994, 2005), held that mental imagery served to depict, not describe, an object, scenario, or phenomenon. This contrasts cognitive scientist and philosopher Zenon Pylyshyn (2002), who argued that mental imagery was experienced in the same way as language, and language is descriptive. Irrespective of which theory is correct, mental imagery might be lost in translation through the process of reasoning. The art of reasoning, from the viewpoint of mental model theory, is defined as 'a conscious and effortful attempt to process the underlying semantics of propositions by the construction of explicit [mental] models' (Markovits, 2014, p. 3). Although mental imagery can be conceptualised in different ways, the consensus by cognitive psychologists is that an individual's capacity to 'contemplate, create, and manipulate mental images' is contingent on the integrity of our brain's structure, its mechanisms and processes.

By extension, I refer to the cultural codebook of organising principles (explained in detail later), comprised of the filtering and classifying processes of the mind (how the mind is organised, if you will). In accord with one's particular cultural codebook, the language of probabilistic-type images in IPCC reports might read like bad syntax and meaningless semantics. As a result, a person's mental image (or mental model) of the climate change problem might translate as incomplete or incomprehensible. Without knowledge of the language of science, interpretation is distorted, making the image illusory. The cinematic-type renderings of the mind are mistaken for real life. They are further distorted as they pass from one social actor to another and from one institution to another. These actors and institutions, to varying degrees, add or detract from the efficacy of the original image as it continues to fade by way of mistranslation, misinterpretation, and/or a framing bias. Not dissimilar to French author and philosopher Jean Baudrillard's (1990, 1994) media concept of simulacra in which copies of copies of an image are rebroadcast by media networks to the point at which meaning is lost. The point being is that images, whether visual or mental, are subject to reinterpretation to the extent that the original image is potentially lost and/or altered along with its original meaning, thus offering only the residues of reality – or a departure from reality altogether.

The concept of simulacra can be likened to framing effects. For instance, post some negotiations, it is the politico-media that for the most part dictates how the final climate change image is framed and hung in the halls of the public forum – a blueprint for thinking and acting and thus a course of action. Arguably, however, the climate change mental model might be modified over time by the publics as new questions arise and through social expectations. Meaning might be lost or gained anew. Alternatively, the dominance of the alpha mental model might increase and become institutionalised as the default model. The default mental model of the climate change object, the slow-moving model that is portrayed as always incomplete is the model intuitively retrieved until a seminal event occurs, such as an abrupt change in the climate, which disproves its efficacy. Ideally, a disposition to update one's mental model in accord with the image described by the science that mirrors newly acquired true belief – knowledge – would prompt an action to avert such an event.

The efficacy of knowledge: its utilitarian and admirable use

'The challenges facing science include ensuring that it lives up to its norms and that its custodians of knowledge are worthy of trust' (Jamieson, 2017, p. 22). Trust is a transient thing. Trust in the communication of scientific knowledge and those who dispense it poses a recurring dilemma for those who use it and act on it. Knowledge is particularly useful when negotiating the frames of reference as depicted in the General Model of the Policymaking Process. Knowledge is grounded in belief. From Plato to Popper the conundrum of verifying what is believed to be true and what is true belief has been one of biblical proportions. According to philosopher Michael Welbourne (2014), provided belief has certain qualities, the foremost being truth, then most people would consider it knowledge. Knowledge, then, is warranted or justified true belief: 'you can only know that P if P is true'. By extension, as Welbourne (2014, p. vii) pointed out, three things must be true:

1. You believe that P is true.
2. P is true.
3. Your belief that P has some especially valuable status-conferring feature like being justified or obtained by a reliable method or whatever – something, at least that excludes or at the very least minimizes [*sic*] the chances of it being wrong.

Although, as Welbourne duly acknowledged, there is no consensus as to what constitutes a valid 'status-conferring feature'. The scientific method is commonly cited. Ultimately, however, validation is left to the cognitive choreography of the user's mind to build his or her world. The mental models that individuals hold are the building blocks that orchestrate and compose that world.

Words, pictures, and symbols are used to create and convey knowledge. How knowledge is used in the General Model of the Policymaking Process, or how it should be used, is not explicit. The rules that govern its use are largely unwritten.

Knowledge used to defend the creation of something useful, such as the climate change adaptation policy of a planned retreat, and to understand something that is complex, such as climatic change, is both utilitarian and admirable. For instance, there is a need to reasonably comprehend the science in order to assess the seriousness of climate change and subsequently respond with a policy equal to the task. But the mere existence of scientific knowledge is of nominal use unless the publics are convinced of its efficacy. For this to happen it must gain endorsement from the court of public opinion.

The 'court of public opinion', subject to pervasive politico-media framing and a diverse array of internally held organising principles – and importantly, what has significance in terms of one's preferred way of life, livelihood, and worldview – will attest to the efficacy of such policy pre- and post-implementation. The mental models formed of the issue of public interest represent a defining blueprint for thinking and acting. Public opinion, however, is not a true gauge of the seriousness of an issue if the publics are ignorant of the knowledge that underpins the scientific theory. Furthermore, as Frank Ruda (2014) pointed out, any action is difficult without knowledge of the situation; true action, that has an influence on and hence transforms the fundamental coordinates of the situation, is more or less impossible. Indeed, the successful design and implementation of a policy preference requires a true understanding of the situation in order to garner an ample measure of public advocacy before it can be transformed into a 'true action'. Theoretically, advocacy in the first instance should be for the science, as a true depiction of the reality that underpins the policy preference – all else follows.

Presuming the normative worldview that good policy is based on fairness and empirical evidence derived from public forums, expert panels, and independent commissions of enquiry, advocacy for the policy is then a judgement of whether it is a commensurate response to the cache of evidence, rather than to the problem itself and/or the 'court of public opinion'. Much to the detriment of the liberal democratic system, the term 'evidence-based policy' has been bandied about like a cliché, often with little reference to the impartiality and credibility of the source. The publics are always super-suspicious of government sources that are referred to by government as impartial or independent. They are made dubious by their appointment in the first instance because of the publics' general distrust of government. Hence the assumption that recommendations made by government-appointed committees are influenced by a disposition to appease and please their benefactor.

In order to hold political pundits to empirical standards of rigour and accuracy, political scientist Philip E. Tetlock (2005) proposed two tests that their political judgements should pass:

> 1. Correspondence tests rooted in empiricism. How well do their private beliefs map onto the publicly observable world? 2. Coherence and process tests rooted in logic. Are their beliefs internally consistent? And do they update those beliefs in response to new evidence?
>
> (pp. 6–7)

Under these conditions, the efficacy of the judgement and also of the policy can be objectively evaluated pre- and post-implementation. Simply stated, if the science is admirably utilised and the publics reasonably comprehend it, a policy preference subsequently becomes public policy and the policymaking process is complete. Theoretically, the optimal outcome has been achieved. Though I question Tetlock's rule that political judgements are consistent with the internally held beliefs of the person making the judgement. Ideally this should be the case; the ideal of what humans and their political organisation 'ought to be' – the *Gemeinschaft* and *Gesellschaft* as proposed by Ferdinand Tönnies ([1887] 2001). But this also contradicts the need for objectivity, which is often inconsistent with rational choice and the divergent organising principles that individuals internally harbour. In this way, the criterion for the internally held beliefs of the person making the judgement to be consistent with a coherent policy judgement is perhaps unrealistic. Furthermore, how is it possible to judge the correctness and efficacy of a judge's judgement based on these criteria? The reality of the everyday is that trust – in judgement – is the fundamental test of our political leaders. But trust in the judgement of our political leaders is fragile and historically short-lived.

In addition, trust in one's judgement is not simply a matter of choice made freely by the individual. Cognitive psychologist Ergon Brunswick's (1952, 1956) 'Lens Model' conceptualised the human condition of judgement making. Brunswick's intent was to capture the behavioural phenomenon of how humans actually respond to the environment's propensity to convey indeterminate information through an array of fallible indicators. He aimed to highlight the subjective nature of psychology as a composite of the subjective nature of the phenomena it studied – in essence, the void between the objective and subjective. Central to his Model was the proposition that cues to events and objects were only probabilistically related to the corresponding actual events and objects. Brunswick separated events and objects of the real world from the cognitive processes of the mind/brain. In between were cues: metaphorically depicted as a lens through which the user endeavoured to construct an accurate image of the real world from the detail provided by the cues. The accuracy of a judgement, then, was probabilistic. A function of the truth in the criterion from the real world as reflected in the validity of the cues. The lens ostensibly represents a mechanism that screens and filters incoming information subject to the user's organising principles, which Brunswick termed 'vicarious mediation'. One's adeptness in judgement is thus a function of one's judgemental lens. Psychospiritualist and psychotherapist Gill Westland (2015) elucidated this concept:

> The brain connects two worlds – the world inside us and the world outside. The brain mediates between these two worlds so that our basic needs are met. Staying alive depends on how successful our brain does this. What is interesting is that it is up to us how we respond (neurologically and physiologically) to the world.
>
> (p. 31)

Following from Brunswick, Joshua Klayman (1988) argued that cues were central to learning in complex and probabilistic environments. Cue discovery was a recursive learning process in which new predictive cues were identified and validated. These cues were then incorporated into the user's mental model of a given object, scenario, or phenomenon, and used to predict some aspect of the environment in relation to them. Cues in this instance are analogous to Johnson-Laird's (1983, 2013) use of premises as discussed in Chapter 3: human reasoning is a process of mental model validation based on the comprehension and degree of truth in premises. Suffice to say, the degree of truth in cues or premises and the subsequent acquisition of relevant knowledge are reflected in the quality of the judgement made.

By extension, the theory of knowledge – epistemology – is supposed to make a distinction between what is known and what is believed to be true: 'Science aims for knowledge, not just belief, and thus cannot avoid issues of truth and justification ... questions of truth and justification lead to empirical research seeking empirical evidence relevant to our theories' (Moshman, 2014, p. 117). To be precise, something that is believed to be true and just happens to turn out that way by chance cannot be conceived as knowledge. Knowledge as a consequence of chance is an intuitive belief, which is a good place to start, but not to end. This is particularly true when discussing predictions of climate change which commonly centre on a broad range of uncertainty probabilities derived by way of the scientific method about particular climatic events occurring (see for example IPCC, 2013, p. 36).

Paradoxically, knowledge of probability theory might provide climate scientists with a safety net of sorts. That is, to avoid making predictions with absolute certainty; or to avoid making predictions that are mistaken for absolute certainty. But that same safety net also provides denialists and sceptics with a semblance of credibility when referencing the probability that the predicted event will not occur at all and thus, the argument that it did indeed occur by chance, and chance alone. This emphasises the challenge of using scientific knowledge appropriately, in a utilitarian and admirable way, as opposed to its duplicitous and/or self-serving framing.

Lasswell (1948) insisted that knowledge was central to democratic policymaking and robust inquiry was requisite in the application of such knowledge to policymaking. He advocated a scientific approach to addressing society's problems by identifying the problem-solving practices that proved most successful:

> Without knowledge, democracy will surely fail. With knowledge, democracy may succeed.... Without science democracy is blind and weak.... If mankind is to adapt civilisation to the ideal of human dignity, each generation must be in a position to profit from past errors and to improve upon past achievement.
>
> (pp. 1, 9, 12)

According to Lasswell the path to democracy is clear ... the universalisation of knowledge with science as its flag bearer. He makes the quest for human dignity a universal ambition and challenges humanity to recant the popular convention that 'if humanity does not learn from history it is destined to repeat it' (Santayana, 1906). The inference is that the collective mental models that populate the General Model of the Policymaking Process are susceptible to linear thought patterns that likely result in circular thinking. A consequence of path-dependency and the mechanistic sequence ostensibly embedded in the Western approach to policymaking that what happens today is simply an extension of yesterday and the day before.

Lasswell also argued that democracy was founded on knowledge – 'knowing' what it is you are voting for is the most fundamental exercise of knowledge. In the case of climate change, adequate knowledge of climate science is fundamental to casting an informed vote for deliberative climate change policy. Otherwise citizens have blind trust in policy practitioners; that they have indeed put aside the politicking and pursued robust inquiry in the formulation of a commensurate policy response.

The excerpt from Lasswell above prompts the question: is 'knowledge itself' sustainable, in terms of its efficacy, if it is ignored or misused by way of duplicitous framing or if humankind does not learn from its knowledge of what works and what does not? Is humankind's crucial relationship with knowledge sustainable if it is arbitrarily misrepresented, forgotten, or dismissed out of hand by those whose responsibility it is to use it in a utilitarian and admirable way? True belief should not be open to discretionary belief. As one of the twentieth century's most influential philosophers Hannah Arendt (2005) wrote, 'The chances of factual truth surviving the onslaught of power [political power] are very slim indeed, it is always in danger of being maneuvered out of the world not only for a time but, potentially, forever' (p. 299).

Educational theorist Richard Smith (2014) said that 'Acquiring knowledge is often less important than learning how to live with knowledge' (p. 35). For instance, there is a large body of empirical scientific knowledge which strongly suggests that urgent action is required on climate change (Garnaut, 2008a, 2011; IPCC, 2014c, 2014d, 2014e; Stern, 2007). Arguably, this knowledge is being misconstrued or even ignored – or some humans are finding it difficult to live with. Josiah Ober (2008), a classical political theorist and historian, has claimed that this wilful ignorance of knowledge is the precursor to a failing democracy. He argued that the closed-door, 'cloistered-expert' process of policymaking (experts appointed by government) often ignored vital knowledge held by the publics. And the translation and transmission of knowledge across all sectors of the publics were fundamental to good policymaking: 'making good use of dispersed knowledge is the original source of democracy's strength' (Ober, 2008, p. 2). For example, is democracy equal to the task of averting climate change without the optimal – utilitarian and admirable – use of climate knowledge? When, from a policymaking perspective, scientific reality is conceivably subordinate to the political currency it affords at the time.

Policy design and implementation: two sides of the same coin

The final sequence in the General Model of the Policymaking Process – policy design and implementation – in the case of climate change policy, hinges on the probability of the event occurring multiplied by the probability of the magnitude of the impact multiplied by the probability of policy outcomes matching expectations. It is arguably the most difficult and complex stage. If indeed it is credible to think of policy design and implementation as a single stage in the policymaking process. On this Lindblom and Woodhouse (1992), and Lindblom (1959), have argued that a consensus on policy goals was unattainable because no single actor possessed sufficient knowledge to make accurate predictions in terms of policy outcomes; therefore, a rational 'step by step' approach to decision-making was of minimal help to policymakers. They further asserted that delineating boundaries between policymaking stages was a fruitless endeavour. This is because policymaking is ongoing and highly interactive with the balance of policymaking power shifting between constituents, interest groups, and political parties. Under these conditions, only incremental gains could be made; thus, Lindblom's reference to 'disjointed incrementalism' at the beginning of this story.

From a more traditional stance, policy design and policy implementation might be viewed as two separate stages – the 'top-down' (policy design) as opposed to the 'bottom-up' (policy implementation) approach (Sandfort & Moulton, 2015). From the top-down approach policy designers are the central actors. From the bottom-up approach, local actors are the agents of implementation and are thus the central actors. Although there is evidence to suggest a conflation of these schools of thought, according to Peter Hupe and Michael Hill (2015) however, policy goals and implementation are still widely viewed as a top-down 'a-political administrative activity'. I suggest that they are two sides of the same coin.

A slightly more radical stance is to consider the analogy of how the brain communicates with its user: '*Communications with the brain* travel *vertically* from the rest of the body – *bottom up and top down*. They *also move horizontally in the brain – left to right and right to left*' (Westland, 2015, p. 32). It is apt to think of knowledge traversing the policymaking maze in such a manner. From the bottom-up and top-down approaches to policymaking; to the left to right and right to left of politics. The policymaking brains trust operates efficiently and effectively – and thus the ecosystem of democracy, as a consequence – wherein the coordination and collaboration of multiple vertical and horizontal feedback loops become rudimentary. Of course an absence of genuine bipartisanship and public advocacy will sabotage this effect.

In Jeffrey L. Pressman and Aaron B. Wildavsky's classic work, *Implementation: How Great Expectations in Washington are Dashed in Oakland or Why It's Amazing That Federal Programs Work at All* (1973), the main argument centred on the disconnect between policy design and policy implementation. They argued that the top-down approach to policy design is based on a set of assumptions about causation between the status quo and the predicted outcome post-policy intervention. If causal linkages are not identified pre-policy design and implementation, then assumptions about causation and thus the theories postulated are likely to be incorrect: 'Implementation, then, is the ability to forge subsequent links in the causal chain so as to obtain the desired results' (Pressman & Wildavsky, 1973, p. xv). In short, identifying and establishing probable cause and effect relationships and embedding that knowledge in the policy design is requisite to successful policy implementation.

Peter J. May (2012) has noted that policy design determines the blueprint for policy execution but might also harbour the seeds to either augment or undermine implementation. In endeavours to augment policy outcomes there is evidence to indicate that countries are now focussing more strongly on the design of public policy (Bentley, 2014). However, it remains that the stated intent and goals of policy are often vague and incoherent which stifles its success. According to political scientist and public policy analyst Michael Howlett (2009), successful policy design requires:

> (1) that policy aims, objectives, and targets be coherent; (2) that implementation preferences, policy tools and tool calibrations should also be consistent; and (3) that policy aims and implementation preferences; policy objectives, and policy tools; and policy targets and tool calibrations, should also be congruent and convergent.
>
> (p. 73)

Prima facie, from a technical perspective, in accord with Howlett foregoing, policy design and implementation should follow a well-rehearsed process of principles and protocols. Provided that government bureaucracies have the cognitive capacity, resources, bipartisanship, and political will to administer and adhere to principle and protocol, good statutory design with clear and unambiguous goals is likely to lead to successful policy implementation and also highlight non-compliance. But this implies perfect political alignment, with perfect and uncontested information at hand, and policy objectives clear and unassailable. In this world, the policy instrument of choice would be apparent and presumably unanimous – which is rarely the case. The disagreement that has dogged the climate change policy debate is testimony to this.

In his endeavours to elucidate this discontent, sociologist Carl May (2013) observed that successful policy implementation is contingent on collective, coordinated, and cooperative social action with the intent to modify social systems in which the actors involved both shape and are shaped by elements of their contexts. Agents of implementation endeavour to negotiate these interactions and curtail confounders that would otherwise undermine the integrity of the policy implementation process. That said, supposedly simple sequences of events in the implementation process depend crucially on complex social interactions governed purposively by the mental models that individuals and interest groups hold. The upside of this approach, however, is that it transfers the onus of success from the government bureaucracy to the civil arm of democracy: if government fails in its responsibilities to implement efficacious climate change policy or live up to climate agreements made, civil society will succeed. That is, if policymakers no longer have the courage and conviction to do what is right then they must step aside. Community clean energy projects, in collaboration with private enterprise, are gaining momentum worldwide and have definitive aspirations to fulfil this role.

Conclusions

Policy is the outcome of a diversity of competing interests, ideas, and ideologies. Policymakers have theories prescriptive of the mental models they hold about how the world 'ought to be'. Politics is central to policymaking and can be defined as the power to realise a preferred outcome even if it does not conform to the general will of the publics one represents.

To illustrate the path from prediction to climate change policy I applied a General Model of the Policymaking Process to the IPCC's *Summary for Policymakers*. Evidenced-based policymaking is about using scientific data to modify human behaviour – in search of the Good Life – by predicting future outcomes. Now more than ever there is a ubiquitous use of visualisation to convey scientific data. This visual knowledge, composed of words, pictures, and probabilities, facilitates the production of visual imagery as a prelude to mental model formation.

Humans form and subsequently hold to particular mental models that prescribe a certain way of thinking and acting. Central to my story is that the mental models held of a particular object, scenario, or phenomenon – climate change for example – supersedes the relevance of political actors and politics, political systems and governance, or whether the school of thought subscribes to a top-down or bottom-up policymaking approach. The outcomes expected and policy objectives set are reflections of the mental models formed and the mental models held – the mental imagery brought to mind either intuitively or reflectively – by the relevant policy designers and agents of implementation.

This mental imagery, however, might be misconstrued and thus lost in translation if one is not disposed to reason and reflect – some measure of probabilistic reasoning is required. In mental model theory, reasoning can be defined as the conscious endeavour to process the underlying semantics of a proposition (in this case the abstract proposition of climate change) by constructing an explicit mental model of its form and function. In this way, the success of policy design and implementation is contingent on the capacity and proclivity of policy practitioners, which invariably comprises the publics, to form a truthful mental model of the climate change problem and subsequently, set clear and commensurate policy objectives and intent. To do this, there is a need to understand the real world and the science that defines it, in terms of the images, maps, graphs, and futuristic simulations conveyed in the IPCC's *Summary for Policymakers*, for example. That is, science literacy and some knowledge of probability theory are more or less requisite to accurately process climate change information.

Lasswell (1948) pointed to science as the fountainhead of knowledge and the foundation of democracy – knowing what it is you are voting for is the most fundamental exercise of knowledge. Like Lasswell, Rawls (2011) had high hopes for a well-educated society and a well-ordered democracy; as well as an enduring faith in the efficacy of knowledge. Understanding the science – science literacy – is necessary in order to make an informed policy judgement. He argued that there is an onus on government to inform and an onus the publics to be informed. This accentuates the imperative of the utilitarian and admirable use of knowledge. Scientific data pertaining to climate change policy design and implementation is entrusted to those to use judiciously not selectively. The way in which knowledge is used is crucial to how we think and act. In this sense, I suspect that Rawlsian democracy is more about the system of thought and subsequent action, and the appropriate use of knowledge as a legacy of that thought and action, than it is about the system of governance itself. If the system of thought and action is deficient, so too is the integrity and dissemination of knowledge emanating from that system, and thus the system of governance.

Here I would make what is essentially a deviation from the main story in reference to the admirable use of knowledge and the means by which it might be obtained. Michel Foucault ([1989] 2012), the celebrated French philosopher, held the worldview that State institutions modify human behaviour by accumulating information on individuals which gives them enormous reach and power. This of course could be construed in many ways – for good or for evil, for example. However, the exploits of Edward Snowden and company reveal to us that an institutional imbalance has occurred. That government has dedicated a significant portion of our resources into monitoring the thinking and acting of our own citizens and the citizens of other countries. The implications of this are yet to be fully realised but the feeling is decidedly totalitarianism and thus an affront to democracy. This of course is another story and as such, is not the place for it here. But in the quest for true belief, it would be a betrayal of the democratic oath if governments imposed a monopoly on knowledge in the name of protecting one's self against one's self.

True belief – knowledge – is derived from cues or premises. Cues can be thought of as a screening mechanism to criterion from the real world as provided by cues (Brunswick, 1952, 1956) – or similarly described as premises by Johnson-Laird (1983, 2013). Cues represent the lens through which the user mediates between his or her world and the world outside. Cues are subject to the personalised cognitive organising principles that users hold. The discovery of new cues is central to learning about complex environments. Learning about complex scientific phenomena like climate change, for example, is contingent on the validity in the cues. Newly discovered cues are incorporated into corresponding mental models to formulate a response to a proposition (a judgement, decision, or solution to a problem).

Mental models are blueprints for thinking and acting. The accuracy of the mental model formed, and the efficacy of a judgement, decision, or solution, is then a probability function of the validity in the cues. Suffice to say there must be a commitment to the validity in the cues or the truth derived from the premises – in this case, climate science knowledge – for either to be of value in the policymaking process. To engage more effortful thinking and reasoning in order to reach a plausible conclusion, a valid mental model, is recognition of this commitment.

Although managing the theoretical, technical, and practical aspects of policy design and implementation, the final phase in the General Model of the Policymaking Process, is arguably the most difficult. Managing the collective mental models held by individuals and the publics is perhaps the most complex and confrontational phase. Disparate strongly held core beliefs that different publics hold have the capacity to destabilise a consensus and undermine deliberative policy action. In short, the crescendo of critique that misconstrued mental models might perpetrate has the potential to derail policy design and implementation. To be sure, if we accept the natural sciences as the foundation of evidence-based policy, they should operate uncensored and free from political correctness and persuasion. If we become subordinate to what is presumed to be politically correct, and selective on that basis, opinion gathers momentum and uncertainty reigns supreme. Evidence suggests that what is politically correct or not is largely a matter of opinion. Correcting deficient mental models that policy practitioners and the publics hold for a given issue, scenario, or phenomenon, is therefore the first, most crucial and lengthy stage in the policymaking process. Doing this in an empathetic and open-minded way should be the primary course of action. A carefully constructed, clear and concise, apolitical language of 'interlistening' is required.

The difficulty, as it stands, is to correct and reconcile the diversity of worldviews personally held by a multiplicity of actors and institutions. This requires exceptional political leadership in both government and civil society. In a liberal democracy, the design and implementation of climate change policy as prescribed in documents such as the IPCC's *Summary for Policymakers*, and public policy per se, is altogether rewarding and costly. Indeed, the probability of re-election, as a matter of rational choice, is firmly indebted to public opinion. Mindful that in a liberal democracy the probability of humankind's survival rests largely on the degree of prudence and prediction with respect to the mental models that policymakers hold, and thus the efficacy of policy made, which is an enormous burden to bear. Paradoxically, the probability of planet Earth surviving humankind might well rest on the collective mental models that humankind holds of the climate change problem. Which, in reality, is not a case of technology innovation but the ultimate test of the human spirit.

5 The Institutions Process Model

Institutions are the gatekeepers of knowledge and a source of competency that empowers or impedes public policy. As such, public policy is inseparable from an institutional context. An institution is defined herein as a cohort of social actors more or less differentiated by a set of collective mental models that prescribe to certain norms and rules that govern their interactions but not necessarily their thinking. To further elucidate this definition, it is necessary to consider institutional theory.

Institutional theory can be traced to German Idealism and a retort to Kantian suppositions of objectivity (Suddaby, 2015). Institutional theory is founded on the premise that institutions and the way in which they are managed are products of social rather than economic demands. Institutional behaviour therefore often contradicts economic rationality. Thus, although institutionalisation contributes to organisational performance and survival by incorporating legitimacy and competence, institutions are path-dependent and characterised by the self-interest and bounded rationality of their members. Moreover, the substance of an institution derives from how it is perceived and branded, and not from empirical reality. For instance, irrespective of an institution's claim to legitimacy and expectations of its competence in accord with that legitimacy, if it is branded untrustworthy, it is also likely to be observed as operating in its own self-interest and by its own rules, which might be real or imagined.

We are all more or less institutionalised. Institutionalisation is commonly viewed as the process by which members come to accept a shared meaning of social reality – viewed as independent from the reality of the individual's thinking and acting but nonetheless accepted as the way things are and the ways things are done (Berger & Luckmann, 1967; Meyer & Rowan, 1977; Zucker, 1977). Once these representations take shape, theories about everything from personality to the economy and the atom (and therefore climate change) are integrated into everyday ways of doing things (Moscovici, 1988). This raises the prospect that the institutionalised mental model of climate change might not necessarily agree with the personally held one; however, it is in the user's self-interest to be compliant in this respect. John W. Meyer and Brian Rowan (1977) explained the institutionalisation process this way:

> Institutions inevitably involve normative obligations but often enter into social life primarily as facts which must be taken into account by actors. Institutionalization [*sic*] involves the processes by which social processes, obligations, or actualities come to take on a rule-like status in social thought and action.
>
> (p. 341)

Contemporary philosopher John Searle (2010) extended on Meyer and Rowan (1977) foregoing. He described facts that exist independent of human institutions as 'brute facts'. Examples of brute facts are the immutable laws of nature, or perhaps the brute fact that everything we perceive is a construction of the brain. In contrast, Searle described 'institutional facts' as those that exist within the confines of a particular human institution but might have only unsubstantiated ties, if any, to scientific reality. For instance, it might be an institutional fact that within some institutions the threat posed by climate change is highly exaggerated, but a brute fact that this particular worldview of external reality is simply a conceptual metaphor (see for example Lakoff & Johnson, 2008). That is, the institution's conception of external reality, in this case climate change, is largely unconscious and automatic, and understood in terms of the common or shared personal experiences that members of the institution hold of it. Brute facts might periodically come to light but are a constant in that they are subject to scientific theory; whereas, institutional facts might bear no other scrutiny than those imposed by their respective institution, even though they might contradict scientific reality.

Classical sociologist Talcott Parsons (1990) argued that to the social actor, actions are viewed in the context of 'a means to an end', which is central to the concept of rationality. Consequently, or perhaps conveniently, if a brute fact dictates that living at the ocean's edge contradicts an individual's chosen way of life, livelihood, or preconceptions, then new 'institutional facts', based more so on heuristics rather than science, are institutionalised. This then is descriptive of Parson's two approaches to the theoretical treatment of institutions: 'the objective' and 'the subjective'. The former is illustrative of the sociological observer who views the behaviour of a social actor as biased, irrational, and prejudiced (see for example Moscovici, 1988). The latter is illustrative of the social actor that views his or her own behaviour as rational (see for example Kahan, 2013). It becomes a choice of the most effective means available to attain a given end. But given that the sociological observer is also more or less institutionalised, a distinction between the subjective and the objective is not so obvious.

Subjectivity is its own unquestioned worldview. For instance, scenarios developed and used in IPCC Assessment Reports are the consequence of inputs, contributed in the main, by numerous climate science experts. Elisabeth Lloyd and Vanessa Schweizer (2014) have argued that since the late 1990s the logic – intuitive logics methodology – that give these scenarios plausibility has been formulated under conditions which render climate change projections as 'very subjective'. Intuitive logics methodology has substantial weaknesses when applied in scientific research because it involves cognitive limitations such as the biases, particularly groupthink biases, and subjective prejudices of experts in the process of forming a consensus of opinion. Note however that we are referring here to a consensus of 'expert' opinion. Whether this narrowing-down would survive the method used is open to question.

Further to this question, as Michael Polanyi ([1958] 2005) wrote in his account of the institution of science and scientific perception: 'The fiduciary passions which induce a confident utterance about facts are personal, because they submit to the facts as universally valid, but when we reflect on this act non-committally its passion is reduced to subjectivity' (p. 319). Polanyi inferred, although does not unequivocally state, that the idiosyncrasies of the scientist inevitably yield to an element of passion and the idea that the truth in facts, the act of knowing, requires an appraisal that is only revealed upon deeper reflection. This then introduces the methodology of intuitive logic – the convergence of objectivity and subjectivity in scientific reasoning – as described by Lloyd and Schweizer (2014) above. The renowned English philosopher, David Hume, framed it this way: 'Reason is, and ought only to be the slave of the passions, and never pretend to any other office than to serve and obey them' (Hume, [1739–1740] 2003, p. 295). Passion dictates our aspirations while reason maps out the path to achieving them. In a crime of passion, for instance, the only role reason plays is how to commit the crime.

The Model

The Institutions Process Model enlists a wide ensemble cast (Figure 5.1). It extends from the General Model of the Policymaking Process (Figure 4.1), and represents a sociological depiction of how the world works. It draws attention to obstacles encountered along the pathway from scientific prediction to climate change policy. To have a chance at achieving a good policy outcome, the challenge is to get from A (prediction) to B (policy) with the integrity and the efficacy of the science and the policy intact. And from the worldview of the policymaker, with his or her job intact.

The Model frames the main institutional actors involved in the quest for preferred policy outcomes. Although it is broadly consistent with the traditional 'institutional approach' whereby government assemblies, legal systems, and bureaucracies are the principal architects of public decision-making and policy outcomes, it is more attuned to public choice theory and thus the works of James M. Buchanan, Gordon Tullock, and Dennis C. Mueller – explicitly, *The Calculus of Consent* (Buchanan & Tullock, 1965) and *Public Choice III* (Mueller, 2003).

Public choice theory purports to the 'rational actor approach' which suggests that policy decisions and outcomes can be explained by the preferences and negotiating power of individual actors – commonly referred to as rational choice. Indeed, the proceeds of institutional democracy are policy preferences, in this case, climate change policy preferences, that survive an intense power struggle between institutionalised actors and their institutions.

As depicted in the Model, scientific discovery is characterised as a 'giant leap for humankind' – from scientific reality (objective knowledge) to public opinion (subjective knowledge). But this is only part of the journey. As epitomised in the age of Galileo and Newton, the acceptance of scientific reality by the institution of public opinion is not guaranteed – to believe or not to believe is a matter of opinion. In a liberal democracy, the relationship between institutions is

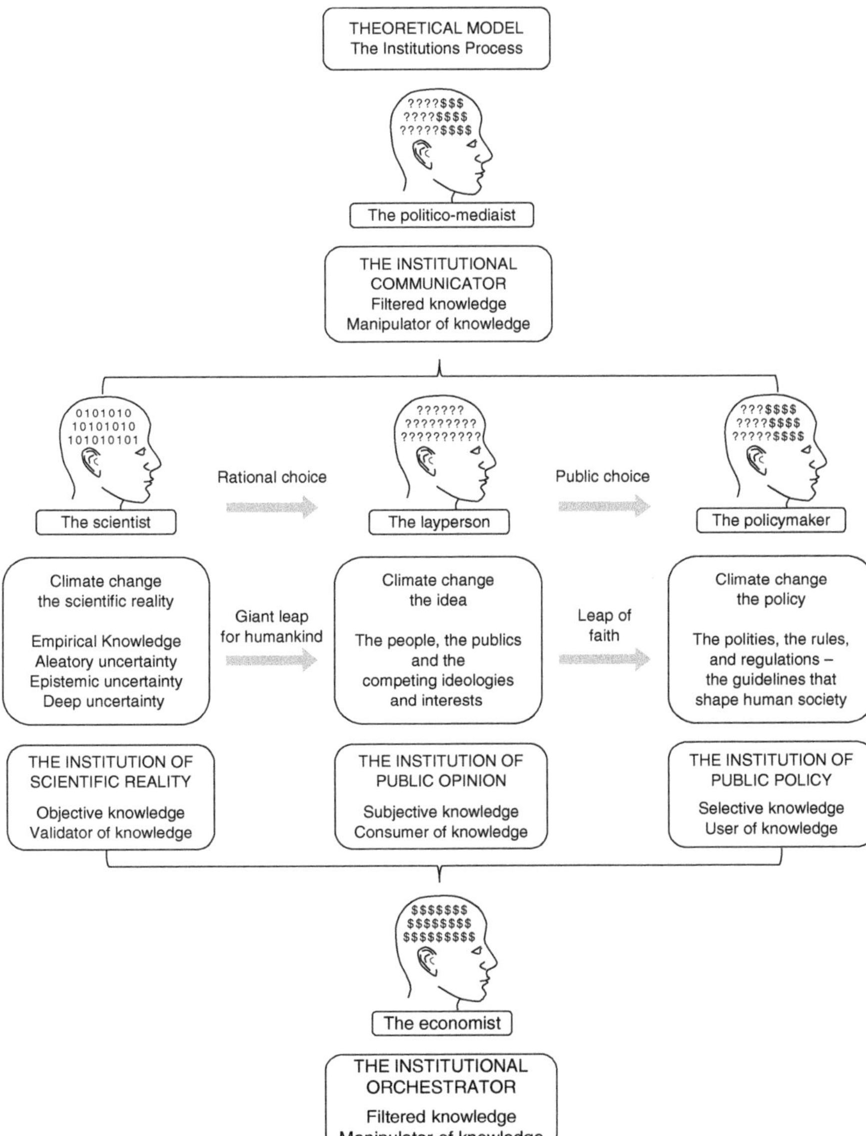

Figure 5.1 The Institutions Process Model.

unavoidable. The relationship between scientific reality and public policy is mediated by public opinion. Public opinion drives public policy. Public policy is founded on fairness and the scientific reality derived from empirical evidence. To achieve realisation, however, public policy necessarily passes through the gates of public opinion as a matter of public debate and in the course of the democratic process. The democratic process is thus subject to the institutional facts held by each institution. For example, although the IPCC (2014b) has recommended that locations that fall within the high-end of predicted climate changes, such as low-lying coastal zones that are vulnerable to sea level rise, require large-scale 'urban climate strategies' such as planned retreat, due to popular opinion, and of course popular demand, we continue to build in these zones and the institution of public policy obliges with the supporting policy.

The Institutions Process Model depicts three primary institutional mental models: the scientist, the layperson, and the policymaker; and two secondary institutional mental models: the politico-mediaist and the economist. They are all signatories to the Social Contract and bound by the obligatory social ties of rational choice and public choice. Each one is a collaboration of mental models that forms an institution. Fragments from all five institutions occupy the aggregated space and are in a constant struggle to impose their respective mental models of climate change, in this instance, as the default mental model. They vie for power over what is true (objective knowledge) and what they believe to be true (subjective knowledge). Either one might achieve ascendancy over the others at a given time.

The Institutions Process Model comprises validators, consumers, users, and manipulators of knowledge. The institution of scientific reality is composed predominantly of validators of knowledge, such as scientists, mathematicians, philosophers, and seekers of truth. It is based on objective knowledge acquired through empirical theories derived from observation and experimentation. The institution of public opinion is composed predominantly of consumers of knowledge, such as members of the general publics, environmentalists, and activists. It is based on subjective knowledge for the reason that 'I believe something to be true'.

The polity, the system of rules that regulates the political process and constitutes the institutional structure of the State, is rooted in the institution of public policy. For this reason, it is composed predominantly of users and accumulators of knowledge, such as accountants, economists, statisticians, policymakers (inclusive of politicians, public servants, and lawyers), the police and defence forces, captains of industry, and religious leaders. The institution of public policy is based on selective knowledge that contributes to the construction of mechanisms to exploit matter and energy and also human resources, and to modify human behaviour.

The politico-mediaist and the economist are considered external, or secondary, policy modellers and less formalised, or perceived as less organised, but are often more powerful in their external capacity to shape policy. The politico-mediaist is a compromise of political and media communicators, and the institutional communicator overseeing the entire political process – reframing and repackaging the actors and the content. The politico-media complex, chaired by the politico-mediaist, is based on filtered knowledge which it implements by dispensing with institutions, actors, and issues deemed unnewsworthy or purposively irrelevant: issues not worthy of press coverage or not in the interests of those that built the network or give it sustenance.

The other secondary institutional mental model, the economist, is instrumental in policymaking, the distribution of resources, and all things financial, and is the institutional orchestrator of proceedings. Indeed, influential British economist John Maynard Keynes (1936) proclaimed that the ideas of economists largely ruled the world. Money, or the utility it provides, remains the fundamental primer and either aids or impedes each institution, but the economist rules as to where it is allocated. Adopting the activities, values, and norms that demonstrate acceptable behaviour within the broader socio-political landscape provide the institution with legitimacy, but money provides it with traction. Subsequently, public and private institutions compete to gain access to economic rent whereby the rules that govern the institutional environment are designed to advantage some but not others (Stiglitz, 2013).

For instance, the genuine pursuit of climate change adaptation or mitigation policies, such as a tax on carbon emissions, could be construed as fundamentally anti-business in the context of the mining and resources industry; thus, not good for campaign contributions that are cardinal to winning elections. Similarly, establishing governmental institutions to independently report on the effects of climate change would be viewed in the same light. In this way, resisting climate change policy recommendations and dispensing subsidies (economic rents) that facilitate the mining and resources industry augurs well with the concept of rational choice and the prosperity of one's institution.

The scientist, the policymaker, and the economist have a fundamental purpose to provide knowledge – a judicious obligation that the layperson is not usually privy to. The layperson, as a member of the institution of public opinion, shapes policy through a majoritarian system. The layperson might also influence policy as an elected member of parliament – as opposed to a career politician – although he or she is then subject to a new set of institutional facts. Significant is that the scientist, the layperson, the policymaker, the politico-mediaist, and the economist are all accountable, and should be held accountable, for averting such perils as dangerous climate change.

The fundamental endeavour of institutions is to predict the future. The scientist, particularly the climate scientist, is involved in predicting externalities in the form of climatic changes. The economist ignores these externalities but presumes perfect knowledge, and has a preference for predicting hard numbers based on cost–benefit analyses and discount rates, and statistics such as inflation and Gross Domestic Product (GDP). The politician is concerned with both the predictions of the scientist and the economist, and endeavours to improve the lives of citizens by directing the economy away from bad outcomes and toward good outcomes. Good outcomes equate to living a healthy and fulfilling life – 'the Good Life for all' – in a prosperous and thriving economy that is not reliant on material wealth more so than a temperate level of sustenance and comfort. It is neither hedonistic consumerism nor an impoverished existence disguised as austerity. We thus arrive at that incessant question of the meaning of life. We might necessarily and constantly reflect on this – why am I doing what I am doing? When the answer fails to materialise, we might even seek the answer from the institution of science. Bear with me if you are not scientifically inclined.

The language of science

Language operates as a 'representational system' which formulises a 'culture of shared understandings' and enables individuals and groups to interpret the world in more or less the same ways (Hall, 1997, p. 1). It is a representation through language that allows the creation of meaning. Meaning legitimises discourse. Discourse – a system of representation that constructs the topics and defines and generates the constituents of knowledge – allows reflection as to how institutions fashion the frameworks of meaning that govern the social structures in which individuals and groups think and act (Carah & Louw, 2015). A fundamental requisite of discourse is effective communication. The word 'communication' derives from the Latin '*commun*' and the suffix '*ie*' – similar to '*fie*' – meaning 'to make or do' (Bohm & Nichol, 2013). Thus, one interpretation is to make something common knowledge by communicating information or knowledge from one person to another as accurately as possible.

Emory A. Griffin (2009) provided a succinct definition of communication: 'Communication is the relational process of creating and interpreting messages that elicit a response' (p. 6). Of course, the response could be precise or obtuse, or silence. Communicating through narratives, whether oral, written, or pictorial, reflects narrative thinking that is a natural, universal, and fundamental capacity of the human mind. This form of thinking differs from paradigmatic or logico-scientific thinking in that the latter works with abstract concepts of construed truth by way of empirical evidence and formal logic that seeks causal relations that lead to truth conditions (Forgas, Vincze, & Laszlo, 2014). More explicitly, the logico-scientific mode (the mode institutionalised in IPCC reports, for example) endeavours to realise the ideal of a prescribed mathematical system of description and explanation using categorisation or conceptualisation and the operations by which categories are established, instantiated, idealised, and related to one another to form a system (Bruner, 2009). Hence, the language of science or of scientists is perhaps not so easily translatable into lay or politico-media terms. Climate change is the archetype example of a failure to adequately communicate a complex scientific phenomenon whose very existence and solution relies heavily on the layperson comprehending it. Blunt warnings, like the one following from Hansen et al. (2016), appear to have only a transitory effect:

> We conclude that the message our climate science delivers to society, policymakers, and the public alike is this: we have a global emergency. Fossil fuel CO_2 emissions should be reduced as rapidly as practical.
>
> (p. 3801)

It is apparent that at this juncture in the 'climate change countdown', as Susanne Moser (2016) has long articulated, 'knowledge itself is insufficiently motivating to take action' (p. 351). Kathleen H. Jamieson (2017) has argued that we need to establish ways to communicate science in a language that is intelligible and precise; framed in a way that neutralises the corrupting influences of heuristics and deception. Jamieson goes further to express some degree of consternation between the language of science and the language of politics, that invariably cross paths:

> Unlike political communication in which selective uses of evidence, categorical assertion, and calculated ambiguity are stocks in trade, science's distinctive norms and structures of enforcement invite forms of communication that take into account the available relevant evidence, specify the level of certainty attached to a claim, and precisely specify the phenomena being analyzed (*sic*) or reported.
>
> (p. 22)

Similarly, James N. Druckman and Arthur Lupia (2017) have argued that science communicators need to choose frames that draw the attention of potential learners while 'staying true' to the underlying research. Scientific findings should be framed to engage diverse publics in a manner that makes a complex issue more personally relevant. Mathew Nisbet (2009) placed the role of scientists in perspective with the contention that if scientists accept the responsibility of determining what is approximately true in the world, then they must also accept the responsibility of communicating that truth in a way that is personally meaningful to their target audience. An explicit understanding of how one's audience thinks and reasons is thus required.

Although public awareness of the climate change debate is presumed to be invariably high, for the most part, public knowledge of the causes and solutions to climate change might be inaccurate by scientific standards (Niebert & Gropengiesser, 2013). Indeed, for those who understand the science, there is a growing sense of despair and hopelessness in the absence of a commensurate action to the threat of accelerating climate change. As Susanne Moser (2016) observed, despite the considerable progress made in the second decade of the twenty-first century, climate change communication needs to transition from a superficial understanding and awareness and concern of the climate change problem, to action. An imperative is to understand how the interaction of social, psychological, and cognitive processes involved in dialogue contribute to sustained engagement that results in measurable shifts in attitudes, opinions, and practice. Bridging the science–practice gap presents a profound challenge. Although extensive research is now available, given the contention that often surrounds climate communication, there is little incentive for scientists to venture beyond academia and engage the publics and practitioners in the actual field.

To gauge the perspective of scientists, John C. Besley and Mathew Nisbet (2013) performed a cross-national comparison to examine the views of scientists from the United Kingdom and the United States toward the public, the media, and the political process. The findings from past studies were added to two large-scale surveys: one from the United Kingdom (2005, $N=1377$) and one from the United States (2009, $N=2535$). The results showed that scientists almost universally believed that the publics' knowledge of science was inadequate and they viewed this knowledge deficit as central to shaping 'risk perceptions, policy preferences, and decisions'. Among other concerns was the publics' scientific illiteracy, which scientists perceived to be a constraint to environmental action, new technologies, and research funding. However, scientists did not exclusively blame the publics for these failings, suggesting that they were misguided and inordinately swayed by bias and sensational media reporting. Paradoxically, this largely negative view of the media was balanced by a positive view of their own participation in media coverage and the belief that the media remained an effective means of communicating with the publics.

Besley and Nisbet further reported that scientists viewed the public as indifferent to becoming more knowledgeable; that members of the public were 'non-rational and unsystematic' in their thinking to the extent that they relied on anecdotes and that they overreacted to minor risks, were self-interested and overly focussed on the sensational. With respect to the public policy debate, scientists believed they had a definitive role to play but emphasised the need for increased public education so non-experts would make policy choices that acquiesced with the preferences of scientists. Scientists also believed that the best method of achieving policy outcomes was through their direct engagement with policymakers. Overall, the analyses revealed few differences between the views held by the scientists from the United Kingdom and the United States.

Of course, climate scientists are not solely responsible for delivering the climate change message. Science journalist Maximilian Schäfer (2011) explained that the aim of the science journalist was to present science research in a way that could be understood by the non-scientist. The science journalist must be able to assess whether the findings are relevant within a particular scientific field and also worthy of the publics' attention. Some knowledge of the subject area is therefore necessary, albeit mindful that the subtleties of original scientific work are generally only fully understood by researchers in the same or a similar field. The main challenge, according to Schäfer, was how to present science in a way that stokes the publics' curiosity and interest without oversimplifying or misrepresenting the facts 'too much'. This invariably leads to conflict between the author and the fact checker. And because scientists unavoidably draw incomplete conclusions stated in probabilities and based on assumptions, the more complex the research, the more difficult it is to convey in a meaningful way. Schäfer noted that climate change exemplified this difficulty.

In a similar vein, broadcast meteorologists are ideally positioned to communicate climate change information to the publics. They are confronted with choices as to what information to omit, what to include, and the degree of technicality in the language used to convey their message. Meldrum, Szymanski, Oches, and Davis (2017) conducted a series of interviews with

American professional broadcast meteorologists that revealed three significant findings: non-negotiable constraints are applied to on-air time, there was constant pressure to be a 'popular' personality, and there was an ever-present trepidation with respect to taking a position perceived to be associated with a controversial issue or political viewpoint. It was evident that these constraints influenced the content and framing of particularly contentious issues like climate change. Without these constraints, it was apparent that broadcast meteorologists would more likely take a stand and present the scientific consensus on climate change. Mindful that the American publics' responses to issues related to climate change can be overwhelmingly positive, or equally negative. Despite these impediments, it was apparent that weather presenters had the potential to assume the role of climate educators.

Climate scientists bear a heavy burden with the fate of the planet partly in their hands. They face a substantial challenge explaining the predicted risks and uncertainties of the potential changes to the climate system to the politico-media, non-scientist, or layperson, and might not have the training to effectively communicate this risk (Fleishman-Mayer & de Bruin, 2014). In the same way, the layperson might misunderstand the science simply because of an incapacity to grasp it, notwithstanding that it might be poorly explained in a language they clearly understand. Nick Pidgeon and Baruch Fischhoff (2014) warned that although much of the publics, and also our political leaders, had recognised the looming implications of climate change, a large majority failed to adequately comprehend the underlying science well enough to assess climate-related proposals and contentions. As Oran Young (2017) duly noted:

> Despite the development of increasingly sophisticated models of the Earth's climate system … we are frequently taken by surprise by the actual behavior (*sic*) of this system, so much so that it remains difficult to educate even the most informed members of the general public regarding the nature of climate change, and so-called climate deniers are still able to gain a certain amount of traction in claiming that the whole issue of climate change is a hoax.
>
> (p. 7)

Extending this, apart from being elected by the majority vote, what credentials do government assemblies actually have which qualifies them to make judgements with respect to climate change policy recommendations? Good judgement is a profoundly arcane thing and politicians are often criticised for a lack of it. Considering this, and the instrumental role that politicians play in the translation of scientific data into lay terms, it is apt to point out that the language of science must not be confused with that of politics (Jamieson, 2017). That is, one must be cautious when deciphering the language of politics when discussing the implications of climate change, for such language is likely to comprise a layperson's understanding of the science imbued with the personal political aspirations and opinions of the communicator. Considering this, should one listen to politicians at all when the subject matter is complex scientific phenomena? Should climate scientists take the lead role in subverting the extraordinary danger of climate change (see for example Stehr & Ruser, 2017)? To be sure, there is an urgent need to overcome the inextricable challenge of communicating climate change in deeply 'polarised and politicised' environments (Moser, 2016). But as Alistair Pennycook (2017) reasoned, 'language is always produced and constrained within a domain of cultural politics … all education, culture and knowledge is political' (pp. 293, 302).

The language of politics

George Orwell's talent for turning political writing into a covert art of satirical imagery was personified in his totalitarian vision reflected in his novel *Nineteen Eighty-Four* (1949). Among other things (see for example Newsinger, 2018), he was a harsh critic of politicians and political language:

> In our time it is broadly true that political writing is bad writing. Where it is not true, it will generally be found that the writer is some kind of rebel, expressing his private opinions, and not a 'party line'. Orthodoxy, of whatever colour, seems to demand a lifeless imitative style.… When one watches some tired hack on the platform mechanically repeating the familiar phrases – *bestial atrocities, iron heel, blood-stained tyranny, free peoples of the world, stand shoulder to shoulder* – one often has a curious feeling that one is not watching a live human being but some kind of dummy: a feeling which suddenly becomes stronger at moments when the light catches the speaker's spectacles and turns them into blank discs which seem to have no eyes behind them.
>
> (Orwell, 1968, pp. 135–136)

Orwell was a storyteller but remained a social and political critic at heart. He has been criticised over the years for his disparagement of the political establishment, its members, and its language. Whether one accepts Orwell's critique foregoing is a matter of choice. As such, it should be evident at this point that language has its persuasive limitations (see for example Nietzsche, [1884] 1974). The language used might be a reflection of the social arrangement one is bound by; and/or, symptomatic of one's worldview (for example Marx and his 'economic structure of society', and Orwell's rather bleak vision of a pending dystopia).

The language of politics is characterised by an individualised and institutionalised set of framing conventions. Thus, the translation of political language into lay terms suffers the subtle nuances to that of the language of science. The competition among mental models is no more rumbustious than in the political arena. It is a contest between the Alpha mental model and the Beta mental model as a measure of mental simulation capacity. The ultimate prize is the internalisation of a particular worldview. *At its bare roots, politics is a battle for the rule of minds – of what future path should be taken, what that path should look like and most importantly, who should lead us down it.* Confounding the latter is that in politics one can never be sure if the worldview expressed in a political communication is the personally held one or the institutionalised one. Thus, the irony that 'fake news' also extends to 'fake politicians' who create 'fake news' but also selectively decry other conflicting news as fake. As reported by editor and political correspondent Eric Pooley (2017) in *Time Magazine*: 'Professional climate change denial is the original fake news' (para. 2). In its purest form, fake news is fundamentally 'bullshit'.

The truth is infinitely flexible. Political speak rests heavily on half-truths and hyperbole, imparting something less than an evidence-based communiqué. Separating a truth from a falsity in a political communication can be a challenge. Arendt (2005) said that 'Truthfulness has never been counted among the political virtues, because it has little indeed to contribute to that change of the world and of circumstances which is among the most legitimate political activities' (p. 307). Arendt was of the opinion that truth is powerless as deceit is powerful; and as such, bullshit is somehow a justifiable tool of the politician's trade. To mask such bullshit and endeavour to make it stick, the politician will occasionally revert to describing it as his or her personal opinion. Truth and opinion are then indistinguishable from one another. But In the case of parliamentary or congressional debate, no endeavour is made to mask the bullshit because it proliferates on both sides of the argument.

'To bullshit' is an exaggerated form of framing. According to Professor Emeritus and philosopher Harry G. Frankfurt (2005) the most striking feature of modern society is the proliferation of bullshit. He emphasised two main traits of a bullshitter: a complete lack of concern with the truth and the intent to mislead listeners by conveying something false as a pretext to pursuing his or her interests. According to James Fredal (2011), bullshit derived from an arrogant disregard for another's worldview in which the hubris of one party, by way of authority or rhetorical sophistry, makes the other party feel undervalued and overlooked. Moreover, Frankfurt (2005) cited the domains of advertising and public relations and the deeply implicated consociation of politics as replete with instances of bullshit. In a normative sense, Frankfurt's description of a 'bullshitter' eloquently befits the politician. With respect to what might be viewed as an archetypal example of bullshit by a politician, it is apt to consider former Australian Prime Minister, Anthony John Abbott's (Tony Abbott) address to the Global Warming Policy Foundation (a London-based group of climate change sceptics) in London in October 2017:

> unadjusted data suggests that the 1930s were actually the warmest decade in the United States and that temperatures in Australia have only increased by 0.3 degrees [Celsius] over the past century, not the 1 degree usually claimed (para. 25).… In most countries, far more people die in cold snaps than in heat waves, so a gradual lift in global temperatures, especially if it's accompanied by more prosperity and more capacity to adapt to change,

> might even be beneficial (para. 27) … at least so far, it's climate change policy that's doing harm; climate change itself is probably doing good; or at least, more good than harm.
>
> (para. 28)

Suffice to say, Tony Abbott is ostensibly a member of the group described as climate change sceptics. As a former Prime Minister, and thus a voice of authority, he is also one of some influence, which does not augur well for the efficacy of a country's climate policy. As Cass Sunstein (2000) noted, like-minded people are inclined to deliberate with each other which, 'because of the limited argument pool', might lead to miscalculations and falsehoods. Further to this, I firmly suspect that group tribalism, whether demonstrated by the discord between right- and left-wing politics, climate change sceptics and advocates for immediate action, or on a larger scale, the United Kingdom's exit from the European Union (Brexit), is the source of much of our anger and angst. Rather than hard core racial discrimination per se, in a globalised world, groups are more and more intent on reaffirming their ideological, religious, ethnic, socioeconomic, national, and demographic identity. And in a world that is so amply interconnected, what better way to do this than by propaganda.

Bullshit can be substituted with propaganda – although propaganda has probably been unfairly stigmatised because of this. What distinguishes propaganda from all other forms of persuasion (with the exception of bullshit) is intent (Taylor, 2003). Its purpose is to convey a message or idea that is devised exclusively to serve the interests of the conveyor. The claim being that all other forms of persuasion are designed to benefit both sender and receiver, or at least purport to do so. Philip M. Taylor's description of propaganda has claims to the sinister side of framing effects. The term propaganda connotes a 'sinister' undertone. He described propaganda as 'a "dirty trick" utilized [*sic*] by "hidden persuaders", "mind manipulators" and "brainwashers" – Orwellian "Big Brothers" who somehow subliminally control our thoughts in order to control our behaviour to serve their interests rather than our own' (Taylor, 2003, p. 1). No less than the rule of minds – the mass social engineering of the mental models that the publics hold.

In accord with the response from scientists (see Santer et al., 2017), an example of political propaganda (with some degree of bullshit interspersed) is apparent in Scott Pruitt's senate testimony post his nomination (and later appointment) as head of the U.S. Environmental Protection Agency (EPA) in early 2017. When questioned about the Earth's warming trend over the past several decades, Pruitt responded with:

> I am aware of a diverse range of conclusions regarding global temperatures, including that over the past two decades satellite data indicates there has been a leveling (*sic*) off of warming, which some scientists refer to as the 'hiatus' … I am also aware that 'warmest year ever' claims from NASA and NOAA are based on minimal temperature differences that fall within the

> margin of error. Finally, I am aware that temperatures have been changing for millions of years that predate the relatively short modern record keeping efforts that began in 1880.
>
> (p. 145)

Parts of Pruitt's testimony that referred to global warming were strongly refuted by scientists. In a direct response to Pruitt's claims, scientists retorted with:

> Satellite temperature measurements do not support the recent claim of a 'leveling (*sic*) off of warming' over the past two decades.... In two out of three recent satellite datasets, the tropospheric warming from 1979 to 2016 is unprecedented relative to internally generated trends on the 38-year timescale.
>
> (Santer et al., 2017, p. 1)

The 'global warming hiatus' that Pruitt referred to, must be referred to in the context of decadal climatic change, and the different datasets used in modelling by different climate scientists. For those who are interested, Medhaug, Stople, Fischer, and Knutti (2017) provided a detailed explanation of the 'global warming hiatus', although the significance of its occurrence remains contentious.

The question of whether the publics are able to detect the bullshit and propaganda dispensed by politicians, and act accordingly by adjusting their voting preferences, remains largely unanswered. Drew Westen (2008) observed that with respect to voters, emotion wins over reason with feelings toward party principles and their respective candidates clearly trumping the specifics of party policy. But in the context of politics, what inspires such impassioned rhetoric? Considering the magnitude of the implications of climate change, such a point of difference between the climate change policies of liberal and conservative governments hardly suggests that it is merely a coincidence, politically orchestrated, or media-inspired. So, to err on the side of rationality, the difference between political parties is generally accepted as a genuine difference in ideology. As journalist and political advisor Lindy Edwards (2013) wrote: 'The great political theorists of history were for the most part ideologists seeking to influence the politics of their day' (p. 11). Ideology provides the coordinates to making meaning out of the paradoxes and propositions of the everyday, but to what end? From the viewpoint of politics, ideology (not unlike culture) is a set of ideas, attitudes, and beliefs that relate to preferences and prescriptions about how to attain the Good Life – but for whom? Is it a case of the Good Life for the like-minded in accord with their ideology; or the Good Life for all, absent of all ideology?

Ideology, once the 'science of ideas', is now associated with erroneous beliefs and the deluded (Goodwin, 2014). Because of this, one might surmise that ideology is the enemy of open-mindedness. Open-mindedness might be viewed as being predisposed to update one's mental model as new knowledge becomes available. Conversely, being 'too' ideological by rigidly adhering to a past prescription in a rapidly changing world can be positively foolhardy. Moreover, in contemporary politics, 'ideological' is often branded a 'dirty word': 'we are deeply suspicious of ideology … it is an attack, a critique, a put-down to describe someone as being ideological' (Edwards, 2013, p. 25). Nonetheless, a political ideologue is a political actor guided by a certain partisan worldview. The fear then, if one was to adjust one's ideology, or move it closer to scientific reality, for instance, one risks alienating oneself from the cast and crew to which one belongs. This is because subscribing to a different reality, or ideology, diminishes the power of the politico-media to coherently propagate its preferred message. One's political narrative becomes garbled and its meaning and intent are fractured. Assuming of course, that someone is still listening.

Deliberation about climate change policy is different to the implementation of deliberative climate change policy. Bullshit or propaganda might be deceiving and demeaning (Fiske, 2014; Frankfurt, 2005; Fredal, 2011; Taylor, 2003) but talk is cheap if it does not translate to an efficacious action and is therefore, 'costless, nonbinding, and nonverifiable' but has the potential to influence an audience's worldview (Farrell & Rabin, 1996). For example, the overall results from the 2016 Global Green Economy Index (GGEI) report (Tamanini, 2016)[1] pointed to the United States and China (the world's two foremost air polluters) as countries in which perceptions of their green economic performance far outweighed their actual performance. Relative to the 80 countries ranked, China was 12th on perception but 64th on performance. The United States was 2nd on perception and 30th on performance. The Nordic countries of Sweden, Norway, Finland, Iceland, and Denmark continued to rank highly, all within the top ten of high performing countries. The GGEI singled out Australia's continued poor performance, ranking it 16th on perception and 55th on performance. This indicated that Australia was very effective (relative to 80 other countries) at appearing to lead on the issue of climate change – through the use of cheap talk or bullshit if you will – but ineffective at matching that rhetoric with action. Money talks and cheap talk is simply worthless, but can prove costly if it is believed.

Notwithstanding cheap talk, political leaders are charged with the formidable task of realising their country's hopes and dreams. Likewise, policymakers are under intense pressure to respond effectively to multifaceted, and at times, paradoxical issues – as a measure of their response, they might find themselves 'damned if they do, and damned if they don't'. Countries hold disparate priorities that lead them to perceive the issues that endanger the Earth (such as climate change) differently. They array these issues to reflect the priorities of their respective countries which make it virtually impossible to agree on a common global agenda. What to do first and how to do it are the main priorities of nation-states.

Central to the system of nation-states is the Westphalian system (see the Peace of Westphalia and the signing thereof in 1648). This system, comprised of the relationships among the nation-states that constitute the system, is based on the principles of sovereignty and non-interference. The government of each sovereign nation-state is recognised as having sole political authority over the people and territory within its sovereign borders. Climate change negotiations, then, in the case of GHG emissions reduction targets, are set against a backdrop of sovereignty, non-interference, and the priorities unique to each sovereign nation-state. From this worldview, 'national interests trump climate-related ones' (Harris, 2015, p. 142). As such, a system of sovereign nation-states cannot cope with issues like climate change that encapsulates the entire planet. I surmise that an entirely different system of world-order is probably needed to guide humanity through complex global events, like climate change, and into a liveable future.

Politicians base their decisions on the costs and benefits they will incur as a consequence of their decisions. As such, their language is a machination of the decisions they make grounded in the supposition that the optimal strategy is to conform or risk exclusion. Dan Kahan (2013) argued that rational choice is a function of 'ideologically motivated cognition' that advances individuals' interests in forming and maintaining beliefs that demonstrate their loyalty to their respective affinity group. This was premised on the rationale that the desire to identify with a cultural group is the prevailing norm when confronted with the cost of taking a position at odds with that group, which could prove disastrous. Rational choice is the dominant tactical approach.

Group membership is desirable and ostracisation is to be avoided, even if it means suppressing or adjusting the default mental model held. Faced with expulsion, party members will commonly choose to 'toe the party line' (Kahan, 2010; Russell, 2014). It follows that a model of like-mindedness, as a product of rational choice, can theoretically produce only one like-minded result. Accepting this model is to accept the terms of institutionalisation. However, as Kahan (2013) pointed out, it is not necessarily the mental model held in a member's conscience that guides his or her institutional behaviour, but the one that provides the most security and comfort through group membership.

Acknowledged here is that in a liberal democracy party members might choose to 'cross the floor'; however, in contemporary politics, this is usually seen as a sign of disunity and treated with disdain, rather than a conscience vote and the opportunity to exercise their democratic rights. A liberal democracy guarantees a perennial conflict between party loyalty and the democratic oath. Party members might choose to bear the mental anguish – cognitive dissonance – because their actions do not match their thinking, in an endeavour to conform to party preferences. The alternative is to adjust their beliefs, or more precisely, their mental models. However, this might appear disingenuous, and places question marks over the conviction of their principles.

The meaning and intent of political language might not be obvious. Forgas et al. (2014) argued that the public are largely unaware of the subtle and complex processes the politico-media uses to influence their judgements and decisions. Studies reveal that political sophistication can alter the way people process and evaluate information about politicians and events (Bertolotti, Catellani, Douglas, & Sutton, 2013). Citizens rarely have direct access to political and economic facts, relying on explanations or debates by a group of actors typically composed of incumbent government officials, members of the opposition, journalists, pundits, and commentators. Moreover, politicians do not simply provide citizens with information but frame it to increase their chances of re-election and to place their opposition in a negative light.

As mentioned above, Kahan (2013) described this as ideologically motivated cognition in which people make rational calculations to demonstrate their loyalty and preserve their status within their chosen group. For instance, it would be unlikely that all adherents of conservative politics are climate change sceptics. For those that do actively endorse the climate change threat, however, it would likely be deemed a betrayal to the party cause and therefore at substantial personal cost if they were to openly express their worldview by censuring their party for an inadequate climate change policy response, should that be the case.

In addition to cherry-picking specific points of reference to support an argument, political debates about the implications of climate change might also be framed counterfactually. Counterfactual communication, prompted by the need to respond to negative or unexpected events, has propaganda potential by allowing political actors to moderate the seriousness of negative outcomes by portraying even worse (or better) scenarios if a different course of action had been taken (Bertolotti et al., 2013; Forgas et al., 2014). In accord with social psychologist Neal Roese (1997) counterfactual thinking is generally activated by a negative effect from a past event (thoughts of what might have been). The use of downward counterfactual comparisons, such as the outcome could have been even worse if a different course of action was taken, is balanced by upward counterfactual comparisons, such as the outcome could have been better if a different course of action was taken. Counterfactual framing proves only one thing – everyone is an expert in hindsight.

Political language laced with vitriol and without substance is unlikely to yield a bipartisan approach to climate change policy; nor is it likely to convince the publics that an intelligent and constructive debate on the issue of climate change has been had. British psychotherapist Andrew Samuels (2015) described how individuals in the West 'agonise' over the collapse of political language, declaring that 'politics is broken'. Apart from Orwell's quite savage critique of the abysmal standard of political verbose at the beginning of this section, much time and energy are spent orchestrating Machiavellian plots to destabilise the opposition.

I shall refer here to the Australian political system as it represents an exceptional case – unmatched in any other country – of contemporary political disarray, factional fighting, and leadership challenges. Although some might contend that the United States, or the United Kingdom, currently have legitimate claims to this. Nonetheless, between 1970 and 2016 Australia had 73 successful leadership coups in State and Federal politics (Tiffen, 2017). There are several possible reasons for this, such as a three-year federal election cycle and compulsory and preferential voting, although no one reason stands out. From the publics' perspective, the reason is most probably a scarcity of good political leadership. Not surprisingly this has had a deleterious effect on progress with respect to important issues, such as climate change. In light of such disarray,

I suspect that it is the civil arm of democracy that has driven Australia's success over the past two decades, not to mention China's unprecedented economic expansion and its demand for Australia's raw materials coinciding with this success.

Through the years of the Howard Government (1996–2007) to the time of writing, in the absence of substantive and deliberative debate, an evasive and abrasive exchange of half-truths over climate change has ensued. Professor of Public Ethics Clive C. Hamilton (2011) reported that the battle over global warming (climate change), a long-running 'culture war' between conservatives and liberals dating from the 1960s, reached fever pitch in 2010–2011. The language of politics descended into death threats and extreme insults (see also Beeson & McDonald, 2013).

What should have been a rich and rewarding discourse on the pressing matter of climate change was reduced to a petty imbroglio over a tax on CO_2 emissions. Well-respected journalist and political commentator Laurie Oakes (2013), described the deteriorating tone among Australian parliamentarians in the period 2010–2013 as becoming nastier and more personal, both inside and outside of parliament. Foreign correspondent and author Nick Bryant (2014) concurred, noting that the growing adolescent behaviour of Australian politicians was only equalled by their malice. Similarly, former federal government minister Barry O. Jones (2012), argued that debates between politicians about issues such as climate change had been deformed by the 'cherry-picking of evidence, denigration of opponents and mere sloganeering', which had led to the 'infantilisation of democracy wherein citizens were deemed incapable of comprehending major issues' (p. 7).

Little can be accomplished in the absence of truth. The efficacy of communication is a function of the degree of truth and coherence in discourse that yields a corresponding mental model of the scenario described. Through discourse a sender describes an object, scenario, or phenomenon, which is translated into a meaningful communication by a receiver. However, the mental model held by the sender and the corresponding mental model formed by the receiver, as a consequence of discourse, might not match. Especially when the message is received through the lens of the receiver's own thoughts and interests – which the receiver tends to defend – irrespective of whether they are true or coherent.

This defensive mechanism is associated with the esoteric concept of self-interest. Ecologist Garrett Hardin, in his work *The Tragedy of the Commons* (1968), argued the disastrous consequences of everyone rationally pursuing their own economic self-interest. If each individual was to maximise his or her return in a world of finite resources there could be only one outcome: 'Ruin is the destination toward which all men rush, each pursuing his own best interest in a society that believes in the freedom of the commons' (Hardin, 1968, p. 1244). Hardin used the example of the 'rational herder' in which all herders grazed their animals on common land to the point of overgrazing and eventual collapse of the common resource and the ecosystem services it provided. In the contemporary world, the same can be applied to the oceans, rivers, and fresh air. Crushed by the clamour of economic self-interest and petty bickering, the fundamental right to a pristine environment is now barely audible in political speak – no less than a 'tragedy of the commons' on a world-class scale.

Rational choice theory

In fairness to the self-absorbed peculiarities of political speak, Karen L. Higgins (2014) suggested that we all live in a bubble, a mental model created by our minds that sees the world narrowly and most often in our own self-interest. Alternatively, the idea of 'the good communitarian' might suggest that each member of a community is looking out for one another in the interest of their own self-preservation. There is strength in solidarity. But the good communitarian is perhaps a redundant hero. One's obligations in a global society now extend well beyond one's sovereign borders. I suspect that the contemporary prototype is the cosmopolitan. The cosmopolitan's self-interest transcends both the human and the non-human worlds. The cosmopolitan vision is in everyone's self-interest.

Ulrich Beck (2010) declared that, to the extent a world public can be awoken from its slumber to the reality that the nation-state system is undermined by global risks, such as the threat of climate change, which bind developed and underdeveloped nations together, something historically new can come to light. It is a cosmopolitan vision whereby people view themselves (their social reality) as both citizens of an endangered world and part of their local histories and survival situations. But in order to attain this collective vision in the face of climate change, how does one protect the rights of individuals to pursue happiness and freedom while at the same time ensure equality and justice for all? A key response to managing these tensions is rational choice theory, a concept that dominated political science in the late twentieth century.

Based on the principles of the liberal tradition, rational choice theory symbolises the importance of rationalism, individualism, and maximising one's life chances; and in particular, limited government intervention in the free market economy. Rational choice theory, however, raises an interesting paradox. James M. Buchanan and Gordon Tullock (1965) posed the question of how in the political sphere – where the implicit assumption is that the pursuit of private gain is ostensibly 'sinful' – the individual is somehow able, when oscillating between the private and the social realities of life, to shift psychological and moral gears from the realm of the private to the realm of the public. Assuming that bureaucrats are equally rational and driven by self-interest, it is difficult to imagine that it would be simply a matter of switching worldviews between self-interest and public interest. One's worldview is normally not so malleable as we have discussed. I would also add that irrespective of whether a decision is made under the pretext of self-interest or public interest, equal constraints are exerted by both on what might be a prudent and viable policy, or in fact the opposite.

It stands to reason that if bureaucrats wish to maximise their self-interest by remaining in office, they will bend to the majority-interest of their electorate guided by the democratic process of majoritarianism. Moreover, the theory of rationalist democracy is premised on the assumption that individual conflicts of interest will be abandoned once the electorate becomes fully informed. The question then remains: what if the electorate simply fails in its duty to be informed and to grasp the urgency or magnitude of the moment?

In this instance, does it remain a case for rationalist democracy or should it be a case of the smartest person in the room? Such a suggestion stokes rumblings of an autocracy or dictatorship. But there is also a case to be made for a requisite level of competence before playing a role in governing one's country. That said, the GFC (2007–2008) confirmed that the smartest person in the room is not nearly as smart as he or she might think, and this included some highly educated people. Alan Greenspan, Chairman of the Federal Reserve from 1987 to 2006 (the lead-up years to the GFC), personified this. Greenspan failed to predict the collapse of the free market and by his own admission acknowledged that he had found a flaw in his model [mental model] of how the world works (Komlos, 2014). Greenspan inadvertently believed that the self-interest of organisations, in particular, the banks, would serve to protect both shareholders and the bank's equity. Greenspan's demise accentuated the fact that economics fails to account for the ideology of the economist. This oversight rendered the mainstream economic worldview as misleading – based ideologically on the way things 'ought to be'. Perhaps as some consolation to Greenspan, recent decades are replete with hundreds of examples of leaders who have fallen prey to their own ideologies and cognitive biases (Thuraisingham, 2013).

Despite a plethora of competing worldviews, a distinction between self-interest and public interest is immaterial if prudence is a guide to good policy. Independent scholar Richard Wilson (2006) observed that in practice and in the case of highly contentious issues, the views of experts, businessmen, politicians, mediaists, lobbyists, quangos, and other pressure groups overwhelm any formal system of rules that could be applied. Climate change is one of those highly contentious issues. Adding to the irony, in the absence of prudent leaders – perhaps mediagenic also – as evidenced by the current state of the world, the downside is comprehensive. But how does one assess the abstract quality of prudence? Wisdom (or prudence) was defined by Russell L. Ackoff (1999), as 'the ability to perceive and evaluate [or mental-model] the long-run consequences of behavior [*sic*]. It is normally associated with a willingness to make short-run sacrifices for the sake of long-run gains' (p. 17). Similarly, according to Sandra Waddock and Patricia H. Werhane (2010), wisdom involves 'seeing', and the wisdom of seeing implies the power to perceive that which is not yet present. No-doubt a prudent politician is a sought-after commodity: or prudence is a sought-after commodity by politicians.

The question is, irrespective of whether climate change policy is made under the pretext of self-interest or public interest, at what point must prudence prevail? For instance Garnaut (2008a) argued that on the balance of probabilities the rational choice is to accept the mainstream science on climate change. Whether this argument derives from rational choice (self-interest) or public choice (public interest) is of no consequence when it comes to the efficacy of the policy response. In order to rationalise a majoritarian system, however, there is a need to define the impetus that underlies the policy preference and the veracity of the premises on which it is based. This brings to the fore the egalitarian notion of public choice theory.

Public choice theory

Public choice is fundamental to the Institutions Process Model (Figure 5.1) because it questions the role of policymakers working in the public interest and highlights the implications of individuals acting in their self-interest in the public forum. Whereas rational choice theory emphasises the role of formal structures, rules, and regulations that guide citizens' behaviour, public choice theory centres on the relationship between government, bureaucracy, and society, and is concerned with explaining how this relationship serves to inform rational political decision-making and the formulation of public policy (Haigh, 2012). Studies of public choice theory concern how people move from individual to group and adjust their ethical frameworks accordingly. A dominant theme is that rational choice also pervades public choice (Buchanan & Tullock, 1965). For example, the political ideologue who is a climate change sceptic would find it difficult to denounce his convictions in the name of public interest. He or she might even conclude that the publics do not know what is in their best interest. That they are victims of exaggerated claims about the dangers of climate change.

Like many terms that comprise the word 'public', a definition of public choice is not immediately apparent. Although James M. Buchanan and Gordon Tullock are recognised as the founders of public choice theory (*The Calculus of Consent*, 1965), Mueller (2003) advanced perhaps the neatest explication. He pointed to Aristotle as the classic fountainhead of the science of politics, and likewise to Adam Smith in the case of the science of economics. As Mueller explained: political science is the study of human behaviour in the public arena while economics is the study of human behaviour in the marketplace. In the past, the fundamental assumption in political science was that political man pursues the public interest. While the assumption in economics was that all men pursue their private interests. Quite simply, public choice theory suggests that political man and economic man are one and the same. Public choice theory can then be defined as: 'the economic study of nonmarket decision making, or simply the application of economics to political science' (Mueller, 2003, p. 1). In this sense, rational choice is also referred to as public choice, social choice, as well as formal choice (Anderson, 2014).

It is a case of when and what rules apply. Public choice theory perceives bureaucracy as a hierarchal framework of institutional rules – mindful that institutional rules necessarily cut across moral and legal boundaries – that compel actors to perform particular types of actions that guide or modify their thinking and acting. Bureaucrats, however, or as one would presume, are like everyone else. Their self-interest has the capacity to override the compulsion to comply with institutional rules. This is evidenced in the formulation of policy that is adversely affected by the competition between government departments. All is fair in politics when it comes to gaining access to funding and control and ownership over policy initiatives. This extends to the lack of bipartisanship on issues of importance wherein politicking takes precedence over policy. It is further demonstrated by bureaucrats' aspirations to control resources, accredit prestige to their position and deploy strategies to safeguard their agency, sometimes at a substantial cost to the publics.

Indeed, political leaders will nurture the government departments that concur with their respective ideologies. Departments that do not acquiesce might lose funding or be made obsolete. For example, a government-sponsored department that produces independent reports on climate change is likely to be curtailed under a government that holds ideological reservations about the scientific reality of the climate change threat. In this way, the self-interest of political leaders imposes itself across the broader community by way of funding and thus ascribed prestige to those departments that conform to their worldview.

Mueller (2003) argued that conventional economics assumes that all persons pursue their self-interest and the fundamental behaviour postulate of public choice is the same as economics, that 'man is an egoistic, rational, utility maximizer [*sic*]' (pp. 1, 2). A rather harsh indictment on the character of man, although Gordon Tullock (1974) was even more to the point. He simply rejected the idea of public interest out of hand to argue that all actors, whether in a public or private setting, act in their own self-interest – thereby relegating culture and history as immaterial to understanding political behaviour. Perhaps, however, economist Arthur T. Denzau and Douglas C. North (1994) offered the most pragmatic word on the subject of self-interest. They declared that, in a politico-economic context, individuals do not actually know what is in their self-interest and thus, cannot act in accordance with it. In reality, people act in response to 'myths, dogmas, ideologies and "half-baked" theories' (Denzau & North, 1994, p. 3) thus confounding the theory of rational choice and humans as rational utility-maximisers.

Keynes (1936) concurred. He wrote that 'animal spirits' prevail as humans swing between spontaneous bouts of pessimism and optimism. Economist and philosopher Amartya K. Sen (1977) went further to describe purely economic man as a 'rational fool' and 'social moron'. The biases of the brain would never allow economic man to conform to such a degree of rational behaviour. More explicitly, Kahneman (2003) referred to this as bounded rationality. Following from the original work on behavioural economics by Herbert A. Simon (1956), rational decision-making and thinking and acting have their limits. They are a function of perception and cognition and thus bounded by, among other things, imperfect information, emotional stress, and the limits of an individual's intelligence. Consequently, rather than the optimising-rational-agent model, 'bounded rationality' was the default mental model that orchestrated economic man's thinking and acting.

There is an argument, however, to suggest that altruism and self-interest are complementary. Although self-interest, whether derived from public choice or rational choice, is largely advocated by Western philosophy and psychology (Baston, 2014), Kurzban, Burton-Chellew, and West (2015) argued that humans are a positively social species and often behave in a way that is costly to them but benefits others. Apart from the benefit that cooperation in groups affords to the individual in terms of safety and subsistence, philosopher Thomas Nagel (1970) posited a simple argument: a disposition to altruism is a condition of rationality. It is recognition of the reality that other humans exist which is a rationale that can hardly be refuted. For example, seeing someone in distress might cause the observer to also feel distressed; helping them can relieve this distress (Baston, 2014). People are willing to volunteer their time, energy, and equipment to search for survivors of extreme climate events, in the aftermath of a hurricane for example, albeit seemingly irrational for self-interested individuals to do so when there is no offer of material reward. Indeed, there are those that subscribe to the worldview that voluntary work is the most rewarding.

This might be construed as self-interest in the sense that helping someone else will relieve one's own stress, which blurs the line between altruism and egoism. As such, it is perhaps more a case of mutual cooperation rather than altruism. Cooperation, or mutualism, implies that both parties benefit and neither incurs a cost. By extension, the cooperative logic of institutions would allow the global society to respond collectively to complex problems like climate change. According to sociologist Richard Sennett (2012), cooperation is embedded in our genes and is the key to sustaining life. It has its roots in a cooperative frame of mind, and thus the capacity to reconcile disparate mental models in matters of public interest. What is in the planet's interest, as opposed to what is in man's economic interest, is a little more esoteric and harder to reconcile.

Indeed, the marketplace is a different place: 'There can be no market economy without the absolute primacy of self-interest [whether labelled rational choice or public choice] to the exclusion of any other motive for action' (Dardot & Laval, 2013, p. 108). As argued by Adam Smith in his *Wealth of Nations* (1776), it is the 'invisible hand' of the free market that guides individuals to act in their own self-interest which results in a greater good for all. Simply put, by manufacturing a particular good with the self-interest of profit in mind, the manufacturer will benefit buyers of his or her good who benefit from its use. Where this credo becomes flimsy is when there is difficulty establishing whether the manufacturer and buyer are also profiting at the planet's expense: expulsing carbon emissions in the manufacturing process, for example.

Irrespective of motivation, institutional democracy should allow individuals to choose whether to actively participate or not. As noted by economic theorist and libertarian idealist, Ludwig von Mises (1998), albeit with an optimistic slant, beyond the quest for material wellbeing:

> Choosing determines all human decisions. In making his choice man chooses not only between various material things and services. All human values are offered for option. All ends and all means, both material and ideal issues, the sublime and the base, the noble and ignoble, are arranged in a single row and subjected to a decision which picks out one thing and sets aside another.
>
> (p. 3)

Could it really be that simple when institutionalisation is such a common condition of modern liberal democracies? The more one is institutionalised, whether imposed through public or private institutions, the less freedom one has to choose. It appears that we have ostensibly chosen to accept the institution of liberalism per se, and its instrument of governance, public policy, as the prevailing orthodoxy. However, if we concur with utilitarian John Stuart Mill ([1859] 1982), to submit to tacit coercion through public policy manipulation was not part of the bargain. Mill argued that a society is only truly free if one enjoys the utility to think and act as one chooses, providing it causes no harm to others. Here, the pursuit of happiness is the principal theme. But knowing in advance what makes us happy – good public policy included – remains a universal conundrum. Political scientist and psychologist, Robert E. Lane (1997), referencing a study that comprised nine different countries, reported that people living in advanced economies and born after 1945 were ten times more likely to suffer from clinical depression than those born before 1945. A truly remarkable revelation, as unwelcome as it is, it tells us that the choices we have made in recent times have not been so good – something appears to have gone askew in the zeitgeist of democracy and the freedom to choose.

According to Nancy MacLean's recent book, *Democracy in Chains: The Deep History of the Radical Right's Stealth Plan for America* (2017), James McGill Buchanan (recognised as the principal founder of public choice theory and winner of the Nobel Memorial Prize for Economics) and Co. are the main culprits. Written much in the mould of a JFK (President John F. Kennedy) conspiracy theory, but instead of 'who killed JFK', MacLean focusses on 'who killed the American dream'. MacLean cited Buchanan as the instigator of a clandestine operation to crush the excesses of democracy through the privatisation and deregulation of state institutions. Moreover, Buchanan executed this plan, which was sponsored by American billionaire Charles Koch, on behalf of his neoliberalist brethren, under the guise of public choice theory.

Reforms, then, as explained by public choice theory, were not reforms at all but were designed to surreptitiously weaken the effectiveness of America's democratic political structure and replace it with an oligarchy under the auspices of the radical right. MacLean claimed that America is now bearing the fruits of Buchanan's blueprint, and the politico-elite are profiting at the expense of the rest accordingly. One might reasonably argue that other parts of the Western world, Britain and Australia for example, have unwittingly or perhaps not, adopted the same blueprint. What is more, in the context of the Institutions Process Model, MacLean's revelations of Buchanan's treachery, should they hold weight, compromise the integrity of the link between the Institution of Public Opinion (the layperson) and the Institution of Public Policy (the policymaker). In essence, paralysing the will of the majority and thus, 'democratically determined goals and needs', as MacLean put it.

Men are not angels and power corrupts

The quest for power and the dogged refusal to relinquish it at any cost is an institution in itself. The publics' worldview of many forms of institutional power is characterised by scepticism. Moreover, there is widespread disenchantment with figures of authority; their motives, actions, and the power they wield. There is a compulsion to not only question their decisions but also their authority to make them. As Frank Furedi (2013) put it: 'questioning authority is [now] interpreted as a duty of a responsible citizen' (p. 153). There is a decided lack of trust in authority – a lack of trust in the authority of religious and political institutions, for instance. Questioning authority is not only the right of every citizen but has become a 'duty of care' for every citizen.

German sociologist and philosopher Max Weber (2009) said that '"power" was the chance [or probability] of a man or of a number of men to realize [*sic*] their own will in a communal action even against the resistance of others who are participating in the action' (p. 180). The misuse of power is closely associated with political corruption. Throughout history, broadly speaking, political corruption – 'activities that threatened to subvert the integrity of public office' – has been accepted as the norm in public office and not the exception (Buchan & Hill, 2014, p. 170). Some dispute remains around a credible and precise definition of 'political corruption' and corruption per se. Mark Philp (2015), a political philosopher and historian, described political corruption this way:

> Corruption in politics occurs where a public official (A), violates the rules and/or norms of office, to the detriment of interests of the public (B) (or some subsection thereof) who is the designated beneficiary of that office, to benefit themselves and a third party (C) who rewards or otherwise incentivises (A) to gain access to goods or services they would not otherwise obtain.
>
> (p. 22)

Corruption per se, however, extends well beyond public office and into every nook and cranny of the private sector. Corruption, plainly, is a failure to conform to some implicit or explicit norm which is driven by an expected gain. Estimates place the cost of corruption (for economic, political, and/or social gain) at more than 5 per cent of global GDP a year (US$2.6 trillion) – a significant proportion of which is bribes worth US$1 trillion (CleanGovBiz, 2014). Official endeavours to combat this have been extensive. Most advanced liberal democracies, among various other political arrangements, are signatories to both the OECD Convention on Combating Bribery of Foreign Public Officials in International Business Transactions (1997), and the United Nations Convention against Corruption (2015). The important global financial hubs of the United Kingdom (London) and the United States (New York) are subject to the UK Bribery Act 2010 and the U.S. Foreign Corrupt Practices Act of 1977 (FCPA), respectively.

Concern about the integrity of public officials and the institutions they administer has a long history. Corruption is commonly associated with bid rigging, fraud, money laundering, and rent-seeking. At the very least it increases the cost of doing business which is usually passed on to those who can least afford it. Corruption undermines the democratic process by demeaning the relevance of voting which is fundamental to democratic representation. Corruption quashes the voice of public opinion and the efficacy of deliberation. The value of promises made by politicians to mitigate climate change, for instance, is of no value if they are not fulfilled. In this sense, reneging on pre-election promises is the most rudimentary form of political corruption. It is a breach of 'trust' and thus a breach of the Social Contract.

There is a normative assumption that society in general is educated in, or has knowledge of, what governance in the public interest is. However, the idea of what is in the publics' interests, whomever they might be, is only relative to who is telling the story – the publics are multifarious with multifarious backgrounds, histories, and interests. One is thus required to reflect on, and reason with some acuity, as to what the actual public interest is, relative to all the publics. We might even extend this to encompass the global publics; and as such, the global threat of dangerous climate change. If public interest is for the 'common good' and Beck's (2010) cosmopolitanism is the order of the day, then we must include the global publics in discussions of public interest in the context of global issues like climate change.

Suffice to say, it is evident that some global publics are in need of assistance. Political scientist Alina Mungiu-Pippidi (2015), observed that 'corruption is associated with massive brain drain' (p. 3). The best educated seek out more meritocratic countries in which to live which subverts their own country's investment in education and stifles their intellectual and economic development accordingly. In certain despotic regimes, but not restricted to them, this might also be a case of 'I think therefore I am persecuted', to borrow from René Descartes. This in turn diminishes the collective potentiality of such countries for sophisticated reasoning and thus broadly, the process of proficient mental-modelling and thus the thinking and acting that it prescribes. A truthful mental model of 'the publics' interest per se' then, is not so clear. That is, to whom do we refer when discussing the 'public interest'? Can we be so myopic that we are only interested in our own specific public interest and all others are of no consequence? For politicians, is public interest a cliché that purports to exonerate them from acting in their own publics' self-interest? For instance, does Western white man think about the indigenous population of his country, or those brought to his country against their will, as part of the general publics?

One would surmise that a common indictment on the 'public interest' model must surely be inequality. Corruption breeds inequality – an unequal distribution of power leads to an unequal distribution of wealth. Inequality can take many forms and is greatly magnified if it is expanded to encompass, for example, the world's publics. Global citizens have rights that extend well beyond their sovereign borders. Even though rich developed countries are largely responsible for past CO_2 emissions, less developed countries and the poor and marginalised will suffer the most from climate change. It is not just a simplistic argument about financial cost but also one of ethics – the 'right thing to do'. Although I suspect that rich countries of the twenty-first century will ultimately bear the financial cost of past digressions as countries with the resources to defend against climate change will be inundated with climate refugees. This will far surpass the refugee diaspora of the early twenty-first century. No country is prepared for what is coming. It is likely to push even the richest countries to the brink of collapse.

Although corruption is more commonly associated with less developed countries, it is not limited to them. Since the late 1980s an increasingly common feature of liberal democracies has been the proliferation of anti-corruption commissions, or 'watch dogs', to scrutinise the public service and those elected to public office. Rarely a day goes by when the media does not report some form of corruption. Political, sporting, religious, and corporate institutions feature regularly. Perhaps, not so surprisingly, many of the world's liberal democracies fall outside the top tier of least perceived corrupt countries. For example, in 2016, neither the United States (18) nor Australia (13) were ranked in the top ten of the least perceived corrupt countries in Transparency International's corruption perceptions index. The United Kingdom just made equal tenth along with Germany and Luxembourg (Transparency International, 2016). In addition, many countries that ranked in the least corrupt category have 'dubious' records in terms of their dealings with other countries – 'dubious' meaning duplicitous. Sweden, for example, was ranked fourth in the perceptions corruption index in 2016; however, the Swedish–Finish company TeliaSonera (37 per cent of which is owned by the Swedish State) has been implicated in bribes to procure business in Uzbekistan which was ranked 156th in the index. It is evident that a kind of institutionalised misconduct exists.

With due respect to this, in the aftermath of a string of corporate malfeasance, the *Sarbanes–Oxley Act of 2002* (SOX) was enacted specific to the honesty and integrity of the United States financial services industry. This had a nominal effect on the thinking and acting of the actors involved in the subprime mortgage scandal in the years that followed. A callous game of betting against the U.S. housing market ensued. Middle-class homeowners would bear the brunt in the first instance – austerity would engulf most of the world in the second. Despite perceptions of man's commitment to honesty and integrity, it is evident that power corrupts. Politicians and bankers are proof of this (Ferguson, 2013; Fiske & Durante, 2014; Stiglitz, 2013). Hence the question: have the publics lost faith in the institutions that serve and govern them? The answer is most probably a resounding 'yes'. Indeed, there is a growing distrust in corporate and political leaders and their institutions which is usually explained away by rational choice, or more plainly, self-interest. But what about the innate desire to do the right thing?

It is apt here to point out that a link ostensibly exists between consciousness and morality (Levin, 2017). Consciousness appears to be a prerequisite for individuals to be morally responsible for their actions. And so, to be conscious of one's wrongdoings but nonetheless carry them out, is criminal – unless one is a zombie. In recent years it has come to light that a corporate set of mental models was immorally programmed to pursue a profit maxim in violation of prudential banking law. Following on from America's subprime mortgage swindle that triggered the 2007–2008 GFC, the LIBOR (London Interbank Offered Rate) scandal stands at the pinnacle (for now anyway) of international banker corruption. The LIBOR is the benchmark interest rate index or interbank borrowing rate that is used as a worldwide reference for a vast array of financial instruments. Estimates of the monetary value linked to LIBOR contracts range between US$300 trillion and US$800 trillion (Ashton & Christophers, 2015). In May 2015 five of the world's largest banks – Barclays Bank, JPMorgan Chase, Citicorp, the Royal Bank of Scotland, and Switzerland's UBS AG – were found guilty of forming a cartel to fraudulently manipulate LIBOR rates and foreign exchange rates to enhance their profits. They established a secretive online chat room to unlawfully collude on LIBOR and Forex Trading. Collectively these five banks, and also the Bank of America that was found guilty of manipulating foreign exchange rates, were fined close to US$6 billion by British and American financial regulators. This was in addition to Germany's Deutsche Bank and the Netherlands' Rabobank that were earlier fined US$2.5 billion and US$1.1 billion respectively, for illegally manipulating LIBOR rates. A report released in early 2018 by the CCP Research Foundation estimated the costs incurred for misconduct, between 2012 and 2016 by 20 of the world's leading banks, amounted to £264 billion (US$368.5 billion). This was an increase of 32 per cent from the period 2008–2012. Despite these findings, criminal prosecutions remain a rarity. Prosecuting individual culpability, it seems, is an onerous task.

James Madison, who wrote the Federalist Papers in 1787–1788 with co-authors Alexander Hamilton and John Jay, that was in many ways a preface to the American Bill of Rights, was indubitably aware of man's follies. Madison noted that government, with its inimitable discretionary powers, would provide the most truthful reflection of human nature. The inference, suitably addressed with an air of diplomacy, was delineated in his reference to angels: 'If men were angels no government would be necessary' (Madison, 1788, p. 337). He elaborated by declaring that because government was administered by men over men, and men were no angels, figuratively speaking, the difficulty in framing government was first to enable it to control the governed, and then somehow instil the obligation to control itself. This goes beyond rational choice versus public choice, but to the excesses of government that Madison believed would lead to the abuse of power if left unfettered.

Some consolation for man's folly can be found in the fallen angel Lucifer, however. Lucifer was banished from Heaven because of a swollen ego and inflated pride to assume the alternate form of Satan. Even angels can suffer from human frailties and be corrupted by power and are thus not exempt from the evils of men – or vice versa as the case might be. In order for the punishment to fit the crime, good must be made distinct from evil; thus, as defined by Alan E. Bernstein (2017), 'the divine artisan uses the devil like an anvil! Satan's temptations shape us to the form God desires' (p. 61). But Satan, like man, shall not be deprived of pity, as portrayed in John Milton's epic poem *Paradise Lost* (1667) as a rather tragic and forlorn figure, suggesting that man has more in common with the fallen angel than just a swollen ego and inflated pride. Ultimately, it seems, the lust for power corrupts both men and angels.

I refer following to an extract from Australia's political scene because it most aptly captures the archetypal mental model of political corruption. In a party-political vein, at the time of writing, the Independent Commission Against Corruption (ICAC; the NSW anti-corruption watchdog established in 1989 in response to the publics' increasing concerns about the integrity of public servants) had unearthed a Pandora's Box of corruption in the New South Wales Labor (progressive left) and Liberal Party (conservative centre right) ranks of government. Professor Emeritus of Political Science John Warhurst (2014) remarked:

> There appears to be acceptance of a way of doing political business that is not just illegal but contrary to high ethical standards … at the heart of the shambles is personal and institutional ethical failure, often driven by the lure of self-interest and advantage, whether it is in getting elected or re-elected or in feathering their own nest.
>
> (paras. 8, 11)

I suspect that many people would agree that the above excerpt accurately reflects the publics' worldview of the generic politician. Of course, it cannot be true for all politicians and their institutions. Likewise, albeit a generalisation, David W. Orr (2014) took aim at public institutions and governments worldwide when he described the current state of political governance as beset with a corruption that had consumed the public interest and sharply diminished the publics' capacity to solve public problems. He argued that over several decades a virulent strain of classic liberalism, formerly used to array against the royal autocracy, had 'hollowed out' Western-style governments and led to enormous budget deficits, the impoverishment of the public sector, the loss of credibility of public institutions, and environmental degradation on a world-class scale.

The illicit inducement through bribes to public officials has at times, thwarted endeavours to protect the natural environment. Bribing bureaucrats has a long history and has become normalised in many parts of the world. Its impact on the natural environment and the powerless is comprehensive. It confounds the stringent enforcement of environmental and energy policy, increases air pollution and deforestation, restricts access to potable water and sanitation, and decreases natural capital and thus ecosystem services (CleanGovBiz, 2014; Fredriksson & Neumayer, 2014; Sethi & Schepers, 2014). It has the capacity to compromise the integrity of institutions whose aim it is to ensure environmental sustainability. And on a global scale, it obstructs the ratification of international environmental protocols.

This is reason to consider moral corruption and the 'duty of care' that governments have to their citizens. If scientific reality pointed to climate change having a devastating impact on human welfare, and governments had the capacity to avert such a danger, they are morally and legally obligated to do so.

And although there is no suggestion of bribery with respect to the recent COP21 held in Paris, France, in November/December 2015, the consensus from several sources was that the aggregate Intended Nationally Determined Contributions (INDCs) by member countries at this meeting would unlikely prevent global mean surface temperatures breaching the dangerous tipping point of 2°C (let alone the aspirational threshold set at 1.5°C) above pre-industrial levels (Boyd, Turner, & Ward, 2015; Lomborg, 2015; Rayner, 2018; Roberts & Pelling, 2018; UNFCCC, 2015). Most member countries have ratified the Paris Agreement but its efficacy is questionable. Perception might create reality but only to the extent that reality allows it to do so. According to Brian Heatley (2017), the Paris Agreement, although considered a success by many commentators (Hilton & Kerr, 2017; Kinley, 2016; Ourbak, 2017), has, in reality, locked the world into an impending abyss:

> Paris [the 2015 Paris Agreement] locks the world into a future where at least a 3–4°C rise by 2100 is virtually inevitable. This will mean a world where there will be massive famine and conflict in much of Africa and the Middle East, serious hunger in South Asia, huge migration pressures but manageable problems in the Americas and more difficult but probably still manageable problems in Europe. Globalisation may collapse, which will be particularly challenging for the UK with its dependence on food imports and international trade.
>
> (p. 10)

More precise modelling by political scientist and author, Bjørn Lomborg (2015), has shown that the aggregate INDCs pledged by member countries at the COP21 were likely to reduce the global mean surface temperature rise by only 0.17°C in 2100. He concluded that even if the COP21 pledges were kept they would not prevent global mean surface temperatures breaching 2°C above pre-industrial levels. Under this premise, governments, as responsible, representative, and impartial institutions, are failing in their 'duty of care' to protect their citizens from dangerous climate change. Governments have the power to avert dangerous climate change. The failure to do so is a corruption of that power and symbolic of institutional failure.

Institutional failure: between civilisation and savagery

> if the misery of our poor be caused not by the laws of nature, but by our institutions, great is our sin.
>
> (Charles Robert Darwin, *The Voyage of the Beagle* ([1839] 2008, p. 812))

The words above, by the English naturalist Charles R. Darwin (1809–1822), showed him to be an exceptional mental-modeller with profound farsightedness. We might reconcile ourselves with the laws of nature, that Darwin referred to, but how do we reconcile the efficacy of the laws made by our institutions with twenty-first-century global maladies like obscene inequality and dangerous climate change? Is the degree of suffering and poverty, relative to prosperity in the world, a fair measure of the failure or success of public policy, and the institutions that formulate it, administer it, and enforce it? Mindful that the laws enacted to enforce public policy amount only to a legitimated threat of punishment of some kind. A financial deterrent by means of a fine and/or incarceration are the standard means. Passive acceptance of state-made law is an expectation under sovereign rule. But laws of the state are not necessarily moral or equitable or even practical. The poor and marginalised bear witness to this.

As the gap between the rich and the poor widens the gap between civilization and savagery narrows. If we were to look back on the present, perhaps in 20 years or so, the most apt characterisation of Western society would be that it was institutionalised, rule-bound, material aplenty, and single-mindedly technologically and economically driven. That the news media was inundated with daily reports of endemic institutional failure – political, economic, and religious – interspersed with the carnage left by global terrorism and the austerity propagated by the politico-elite's hedonistic proclivities. As futurist Arthur C. Clarke ([1962] 2013) lamented: 'the time will come when most of our present controversies on these matters [of politics, economics, and religion] will seem as trivial, or as meaningless, as the theological debates in which the keenest minds of the Middle Ages dissipated their energies' (p. 1). I suspect, however, that most of all we will lament the planetary devastation caused by climate change.

Fritjof Capra and Pier Luigi Luisi (2014) observed that the problem of climate change, as well as energy, the environment, and food and financial security, are simply different attributes of a single crisis, that of perception. This derived from the fact that modern society, and in particular large social institutions, subscribed to a worldview that was anachronistic and, therefore, inadequate to deal with an overpopulated and interconnected world. In essence, our institutions had failed to evolve with the times and were now out of touch with the real world. English sociologist and political scientist Colin Crouch (2004), argued along similar lines. He declared that institutions, central to democratic representation, had failed to purposively fulfil their roles since the early 1970s. The 'democratic moment' experienced by the majority of advanced capitalist countries had reached its zenith during the middle decades of the twentieth century and had since capitulated. Post-democracy was a Dark Age of frustration, boredom, and disillusionment in which powerful minority interests actively made the policy process work for them, much to the exclusion of the general publics.

Crouch observed that a pivotal shift occurred, from democracy to post-democracy, at that defining moment when politico-elites learned too well how to manage and manipulate the mental models of popular demand. They were enabled and emboldened by a drive for privatisation and deregulation in the wake of a neoliberalist constitution that led to the proliferation of domestic and global institutions of immense bargaining power. This culminated in a market anomaly in which some global institutions were 'too big to fail'. Notably, 63 of the world's 100 largest economic entities are now corporations, not countries (Young & Dhanda, 2013). Is this what is meant by globalisation?

Is it what Hedley Bull and Adam Watson envisioned in their classic work *The Expansion of International Society* (1984). Fuelled by decolonisation post-1945 and a communications revolution that gathered pace through the 1990s, the dark side of their vision, of political injustices and economic exploitation, has been largely borne out. Once seen as the coming together of an order of sovereign nation-states, of international and supranational institutions and their transnational political actors – of their values and interests – for the purpose of a safe and secure 'international society'. It had aspirations of the Good Life for all, but also inadvertently enabled the 'proliferation of cybercrime, organized criminal activity, green crime, human trafficking, and terrorism' (Schweppe & Walters, 2016, p. 2). I would add to this list: profit shifters, money launderers, tax evaders and avoiders, labour and resource exploiters, people enslavers, and pornography peddlers. Not quite global anarchy, but what rule of law, formal, moral, or otherwise, do the foregoing entities abide by? Globalisation has created a permissive environment virtually free of any international adjudicator to rule on the malfeasance of global enterprise (Klabbers, 2008). Few checks and balances are in place to ensure the operational efficacy of international institutional law. It is now credible to speak in terms of a globalised culture of crime and corruption.

This of course accentuates the dark side of human nature; perhaps the one described by Hobbes ([1651] 1981) as life in the 'state of nature', which also accentuates the irony of it all. Life in the state of nature is one in which there is no political authority and as such, no law and therefore no-one to enforce the law. Hobbes quite famously described life in the state of nature as 'solitary, poor, nasty, brutish, and short' (p. 186). Think of William Golding's 1954 novel *Lord of the Flies* in which a group of British schoolboys are stranded on a deserted tropical island. At first they enjoy the freedom – in the absence of their parents' sovereignty – but soon struggle to govern themselves and are eventually reduced to lawlessness and savagery. Hobbes advocated absolute sovereignty lest we turn to savagery. There is no global sovereign that has authority to effectively adjudicate over the affairs of international businesses.

The recent 'Brexit' – the United Kingdom's exit from the European Union in the summer of 2016 – was perhaps the first sign that globalisation had run its course and turned full circle. Sovereignty found then lost then found again. Apart from the obvious geo-politicking and genuine disenchantment, three things are apparent here: sovereignty is one of those things that is difficult to live with but perhaps more difficult to live without; the socialisation of diverse societies into a world polity and the promise of a world democracy do not always good bedfellows make, let alone good neighbours; and the scope of good governance has its limits and if we press for too much we might return to a life of savagery. The savagery of a world in which the sole purpose is to acquire wealth becomes the human singularity.

The global movement of capital and labour is tied to the manufacture of goods that are now vigorously traded across the planet. This production, distribution, and consumption cycle is responsible for a growing share of total anthropogenic CO_2 emissions (IPCC, 2014). Indeed, in the race to sign free-trade agreements, when a globalised mental model is required to tackle climate change, privatisation and deregulation incorporated under the competitive banner of Neoliberal Inc., cannot hope to achieve the like-mindedness necessary. The global community is 'minds apart'. In fairness to the task at hand, anyone who has experienced the difficulty in achieving a consensus with respect to a complex issue in a room full of disparate mental models will undoubtedly appreciate the challenge of reconciling a group of 197 sovereign nation-states to like-mindedness on the issue of climate change policy. The concept of a 'free market' has been empowered to the extent that any intervention that might thwart its empowerment is commonly branded as anti-business.

What is a 'free market' anyway? It is a classic misnomer in the purest sense. A fundamental condition of a competitive market is that it comprises countless sellers and buyers of an undifferentiated product so no one person, group, or entity can influence the market price. Although some markets more closely meet this condition, inevitably, monopolies and cartels form. An archetype of this is the Organisation of the Petroleum Exporting Countries (OPEC). The supply and price of oil, an indispensable component of most economies, is more or less controlled by OPEC. OPEC consists of major oil-producers that effectively manipulate the price of oil for political and economic gain. It is 'the world's most successful cartel' (Griffin & Steele, 2013). Cartels fly in the face of a free market. Furthermore, the concept of 'free market' hardly makes sense if one applies the 'polluter pays principle' to oil, gas, and coal producers. These large multinationals are subsidised – both by way of billions of dollars in monetary subsidies (rents) and of favourable policy/tax treatment that effectively sponsor negative externalities – for the pollution they create as a consequence of production.

A report by Bast, Doukas, Pickard, van der Burg, and Whitely (2015), sponsored by Oil Change International and the United Kingdom think tank Overseas Development Institution, has estimated that global fossil fuel subsidies (dedicated to fossil fuel production) totalled US$775 billion in 2012. Despite a pledge by the Group of 20 (G20) member countries to phase out fossil fuel subsidies, recent estimates indicated that G20 countries are providing US$452 billion in fossil fuel subsidies annually. Australia, a major fossil fuel exporter, subscribed to an average annual subsidy commitment to fossil fuel production (2013–2014) of US$5.032 billion. This amount, however, was dwarfed by Russia US$22.812 billion; the United States US$20.491 billion; and the United Kingdom US$9.047 billion. It is evident that G20 countries are knowingly sponsoring the fossil fuel industry and thus undermining their own emissions reduction and climate change mitigation programs.

Inequitable interventionist policy in the free market is ubiquitous. Rent-seeking, for example, creates a negative externality that weighs heavily on productivity. It fosters the expectation that influence, rather than merit, is the most fruitful road to prosperity. Complicit with rent-seeking and perhaps the epitome of the free market institution and the doctrine of laissez-faire (French for 'let the people do as they please'), is the offering of gifts to political parties. Political fundraising has been associated with soliciting prohibited monetary donations or 'gifts', from property developers, amongst others, with the inference that government might treat their development applications more favourably. As French sociologist Marcel Mauss (1990) pointed out, there is an unspoken contract of reciprocity as 'the objects [gifts in the form of money or other items of value] are never completely separated from the men who exchange them' (p. 31). I would add that it is evident that independent strategic planning is needed to correct this failure. This requires a genuine separation between politics and planning – a more substantive role for civil society in 'Earth system governance' – which is idealistic but not impossible. After all, idealism, throughout history, has been a constant influence on reality: the reality of neoliberalism and free-market fundamentalism stemming from libertarian ideals, for instance.

Although writing in the early 1940s, Karl Polanyi ([1944] 2001), an outspoken critic of globalisation and free-market fundamentalism, offered a timeless depiction of the free market economy: 'To allow the market mechanism to be sole director of the fate of human beings and their natural environment indeed, even of the amount and use of purchasing power, would result in the demolition of society' (p. 76). And as Stiglitz (2001) recounted: 'Today, there is no respectable intellectual support for the proposition that markets, by themselves, lead to efficient, let alone equitable outcomes' (p. vii). There are at least three unsatisfactory outcomes implied here: the inefficiency of a free market enterprise, the inequality it creates, and the destruction it causes. To be sure, climate change is an outcome of free market fundamentalism – the freedom to use the atmosphere as a waste disposal dump – free of charge. No less than environmental terrorism if one considers the outcome. Inequality is also pertinent here with 1 per cent of the world's population now holding more than half its wealth. This was not part of the so-called 'Social Contract'. Prosperity depends heavily on social cohesion which rests heavily on the integrity of the Social Contract. A society alienated by extremes of poverty and wealth cannot be considered a just society. *Prima facie*, a society treated unjustly renders the Social Contract null and void.

Acknowledged is that a free market economy (or capitalist economy) is not actually free. Colluders, monopolies, duopolies, and cartels endeavour to corner the free market and carve it up into what they consider to be the most profitable pie. Moreover, a true free market system would not comply with the tenets of pure capitalism if there were the redistribution of wealth from the rich to the poor. Ironically, the free-market system that the world is most familiar with is the one in which, according to Stiglitz (2013), wealth is redistributed from the poor to the rich. As such, although some free market systems might come close to being free, there is usually some degree of intervention. Government intervention is expected and required when the free market fails to function efficiently. But who decides on the latitude between free market efficiency and intervention – one is certainly not led by Adam Smith's 'invisible hand'? Perhaps ironically, instead of the efficient allocation of resources, Adam Smith's (1776) 'invisible hand' might well steal away with Earth's most precious commodity – its capacity to sustain life. What might be good for oneself and one's country might be devastating for one's planet.

At what point does the state intervene to literally save our planet? Moreover, how does one reconcile a pluralistic liberal democracy with the authority of the state to intervene at its discretion? One can invoke the Social Contract. A breach of the Social Contract by an incumbent government could result in that government losing office at the next election. But what if all alternative governments breach the same contract – the contract of a duty of care to protect citizens from dangerous climate change, for example? Which brings us to the next question: who is the Social Contract actually with, or between, and how is it rescinded if it is breached? If the contract is with the ruling incumbent, for instance, as long as there is a ruling incumbent, it can never be rescinded. Under this condition it is not a contract.

The basis of a bilateral contract – between the people and the government – is a promise for a promise. That is, parties to the contract must fulfil their particular promise, whatever that might be. Whether one refers to Thomas Hobbes, John Locke, or Jean-Jacques Rousseau, the idea of the Social Contract purports to the promise to give up certain liberties and respect the liberties of others, and consent to the rule of law as enacted by a Sovereign (law-making and law-enforcing authority). In return, the Sovereign promises that the people will no longer 'live in a state of nature' but a civil society. A civil society, one that is liberated from a life of savagery, is one founded on justice, fairness, individual rights, and the right to good governance. Moreover, the right to do as one pleases as long as no harm is done to others. It is a paternalistic construct but one that is indubitably preferable to a life of savagery. The problem with this is the obvious one: the assumption that the encumbered government, or any alternative government, is just, fair, and respects the rights of individuals. The point being, if successive governments fail to fulfil their promise to the people, the status quo remains irrespective of the Social Contract, which the people are forced to sign and abide by, by way of a law that exacts a punishment of some kind.

Switzerland's Direct Democracy endeavours to address this predicament by holding frequent referendums with respect to modifying laws and changes to the constitution. Switzerland is an archetype of political stability and self-determination. Mindful that Direct Democracy is not openly endorsed by most liberal democracies because it moves power from the government to the people. A common excuse is that referendums are costly and time-consuming, and representative government is a cornerstone of the democratic system, but are still sporadically held if it is in the government's interest to do so. A further point to add is that a liberal democracy is founded on majoritarianism, the will of the majority – the 'general will' (*volonté générale*) as Rousseau put it. There are problems with this: first, in contemporary liberal democracies, governments are commonly elected with only 40 to 50 per cent of the popular vote, which is not a majority (Haworth, 2012). And contingent on the idiosyncrasies of the electoral system adopted, could be much lower. Notably, one can win the popular vote but lose the election, as in the recent 2016 U.S. presidential election. Second, 'representative government' further diminishes majoritarianism: a small body of representatives of the people making decisions on behalf of the people they represent – the entire population. And ironically, history tells us that in order for humanity to progress, we must think and act beyond the majority, even if we are persecuted for it.

Scottish historian Niall Ferguson in his book *The Great Degeneration: How Institutions Decay and Economies Die* (2013), pointed out that institutions determined modern historical outcomes and were therefore vital to Western society's previous success. He argued that for the most part, the failure of institutions could be traced to the dissolution of voluntary associations of civil society that operate outside the realm of public institutions. He cited the gradual breakdown of institutions established by volunteers. These are distinct from the economic and political institutions established and operated by politico-elites because they work explicitly to strengthen the fabric of society. Similar to Crouch's (2004) argument, Ferguson (2013) held that societies start out with good institutions that gradually degenerate into bad ones. He believed that the Western world is now fated to a state of 'institutional degeneration'. Monetary and fiscal irresponsibility has led to mounting government debt that had rescinded the Social Contract between generations. But more regulation of institutions, particularly financial institutions, was not the answer. Ferguson (2013) maintained that punishing the guilty to fit the crime was, and the archenemy of the rule of law was bad law.

Both Ferguson (2013) and Stiglitz (2013) pointed to elite rent-seeking by organisations and interest groups as dominating the political process and undermining the integrity of public institutions. According to environmental economist John C. V. Pezzey (2014): 'If elites are too powerful, they will block new technologies, so as to keep their powers to extract rents from the rest of society, and the nation will then fail (to grow sustainably)' (p. 329). Which brings to mind the endeavours of political leaders to sustain the fossil fuel industry and procrastinate on clean energy technology. A point to add, somewhat ironically, is that central to this story is climate change. And climate change policy serves the interests of policymakers who use rents to shore up their alliances, and also the rent-seekers themselves who benefit from subsidies, grandfathering, mandated markets, and tax breaks.

Stiglitz (2013) maintained that bankers, and the political leaders that appear to support them, had designed a financial system that engaged in excessive risk-taking and market manipulation at the expense of the publics. If things went awry the publics were there to bail them out. The consequence of such institutional failure was to move wealth from the bottom to the top, mainly through rent-seeking. This explains growing inequality and the widening gap between the rich and the poor, and why the gap continued to expand post the GFC of 2007–2008. In the aftermath of the GFC, in order to stimulate and resuscitate an ailing global economy, massive amounts of capital were pumped into economies around the world, referred to as quantitative easing, which continues to this day. A good portion of this capital has found its way into the coffers of the elite.

A report by Oxfam (2016) calculated that the richest 1 per cent of the world's population possessed more wealth than the rest combined.[2] Moreover, a global web of tax havens allowed them to hide (from their respective tax collection agencies) US$7.6 trillion. According to Oxfam (2016):

> The wealth of the richest 62 people has risen by 44% in the five years since 2010 – that's an increase of more than half a trillion dollars (542 billion), to 1.76 trillion. Meanwhile, the wealth of the bottom half fell by just over a trillion dollars in the same period [United States of America dollars].
>
> (p. 2)

Contrary to Ferguson (2013), Stiglitz (2013) emphasised the need for better regulatory policy, but concurred with Ferguson that the rule of law had been compromised. He argued that America's economic system (and in general, advanced Western democracies) were 'captured by moneyed interests' and that markets were not self-regulating: 'Unfettered markets do not work well' (Stiglitz, 2013, p. li). Stiglitz argued that politics shape the market and a 'vicious nexus' had emerged between politics and economics. Indeed, the capitalist economic system, apart from the '1 per cent', had failed everyone.

An example of institutional and capitalist market failure was apparent in the aftermath of the 2007–2008 GFC that caused the greatest recession since the Great Depression of the 1930s. Financial institutions and their executives made substantial profits and bonuses selling 'toxic assets' mainly in the form of subprime loans. Although there was ample evidence to indicate fraud no major prosecutions were made, suggesting that the legal system had been captured by large financial institutions. Pontell, Black, and Geis (2014) described this as 'too big to fail' and 'too powerful to jail'. Ferguson (2013) called it 'the rule of lawyers'. Stiglitz (2013) declared that the once proud proclamation of 'justice for all' had been replaced by 'justice for those who could afford it'.

Institution degeneration and failure purports to the incapacity of institutions to act as intended: cooperatively, decisively, and in a responsible manner. But politico-elites and public institutions cannot be the sole conspirators to the lack of public outrage and often enervated response to climate change. Indeed, the same can be said for the muted response to inequality and poverty. Certainly, the institution of democracy provides a platform for the publics to participate by validating their policy preferences through the democratic vote. But at the same time it allows them only marginal input to the formulation of such policy and the policy choices on offer. These are, in the main, preferred politico-elite initiatives not public initiatives. They are the 'signature policies' of government leaders seeking a crowning legacy of their term in office. Indeed, using carefully crafted filters and frames, 'spin doctors' work tirelessly to capture the minds of the publics and provide the foundations for favourable mental model formation. Despite this sophisticated, but often inept politico-media 'sleight of hand', mental models from all walks of life continue to compete for precedence and prominence. Throughout history, the most celebrated competition is founded on and fuelled by the disparity, conviction, and creativity of the mental models that individuals hold.

There is a much darker side to institutional failure, however. Since all humans are institutionalised to some degree, institutions can be considered as representing the rudimentary mental model for good human behaviour. Post a stream of disturbing media reports regarding child sexual abuse by Catholic priests in the United States in the late 1980s, Jason Berry published the book *Lead Us Not into Temptation* (1992). Berry documented the cases of hundreds of Catholic priests implicated in the sexual abuse of young people. But the breadth and depth of child sexual abuse uncovered worldwide came as a shock to both believers and non-believers. The movie *Spotlight* (Singer & McCarthy, 2015) directed by Tom McCarthy, further shone the light on child sexual molestation in the United States. The movie was based on investigations into the Archdiocese of Boston by the Boston Globe's Spotlight team of journalists. They uncovered a virulent strain of institutional child sexual abuse and a programme of silencing the victims and their families. The institutionalised abuse of power perpetrated by the Catholic Church has left millions of broken lives across the world. A deficient mental model was more or less accepted collectively and embedded deep within the culture of the Church. The fact that it was for a long time dismissed – or passively consented to or ignored – by bureaucracy, is testimony to the fact that any institution, political or otherwise, could suffer the fate of collective mental model failure.

This very ugly chapter in human history is ongoing. At the time of writing many countries around the world were in the throes of investigating child sexual abuse. For example, a Royal Commission was well underway in Australia (Royal Commission into Institutional Responses to Child Sexual Abuse). Under investigation were members and representatives of institutions, inclusive of public and private institutions, such as 'child-care, cultural, educational, religious, sporting and other institutions' (Royal Commission into Institutional Responses to Child Sexual Abuse, 2016, para. 5). The Commission has continued to uncover evidence of endemic child sexual abuse and what appeared to be deliberate attempts by various persons to 'cover up' violations and protect the perpetrators. Most recently, Cardinal George Pell (the Vatican's third most senior official and Australia's most senior Catholic cleric) was recalled to Australia from the Vatican to face charges of historical sex offences. Presumed innocent at this time but the inference is damning.

One wonders how all this began. Was it the Edict of 1910 in which Pope Pius X declared that seven years of age (presumably the 'age of reason') was appropriate for one to confess one's sins? Was it this edict that unassumingly allowed the devious exploits of the few to enter God's house? Or was it more sinister? The best place for purgatory to hide is the last place one would look. Because religion is so powerful and thus influential, should it be only a place for 'grown ups'? Whatever the inception, this was a case of the inculcation of an insidious set of 'institutional facts' into presumably trusted religious institutions. An instance in which self-examination was absent or abstained from, and highly deficient and corrupted mental models ruled essentially unchallenged – a phenomena of institutionalised treachery.

I cannot complete this section without making some reference to the failure of the institution of education. To be sure, sexual harassment is prevalent from Hollywood to Westminster (the seat of British Parliament), and also in our universities, which reflects poorly on their administrators, but is probably a fair reflection of their capacity to fulfil this role. But this is another story and beyond the scope of the one herein. I have discussed in Chapter 4 political intervention in the school curriculum, which I contend to be more extensive than the publics of Western democracies, and probably of most other political systems, realise. Certainly, as it relates to contentious issues of past atrocities and injustices committed, and present ones such as climate change.

Apart from endeavours to somehow reconcile the shameful acts of Catholic priests and others that have engaged in institutionally sanctioned paedophilia, the institution of education has comprehensive accountabilities. It is that place in which all members of society should be free to discuss its direction and the prudence of its methodology. As political scientist Amy Guttmann (1987) wrote: 'What is the democratic ideal that complements democratic education? A guiding principle of liberal democracy is reciprocity among free and equal individuals: citizens and their accountable representatives owe one another justifications for the laws that collectively bind them' (p. xii).

I submit that this 'justification' is often difficult to ascertain with respect to the institution of education. An abstruse rationale or even a culture of dismissiveness, by those in authority, is perhaps more familiar. Mutually binding laws must, at the bare minimum, be open to criticism and question as to their efficacy and moral justification. *Authority must be periodically challenged lest it becomes authoritarianism.* The diminution of democracy through the deprivation and/or distortion of knowledge, and the inequities that authoritarianism invites, weakens individual and civic resolve to defend against unjust rulings and practices that lack transparency. Likewise, with respect to poorly constructed policy and law. The generic practice of which extends outwardly and further to policy paralysis, policy reversal, and the duplicitous tweaking of policy, when deliberative action is duly needed. Climate change policy is a case in point – in many ways a reflection of what we teach and how we teach it. And it matters not whether this pertains to institutional rules or the formal rule of law – democracy should not be selective.

A well-rounded education for all – say, schooled in the arts and humanities and adept at reading, writing, spelling, science, and mathematics – is perhaps a thing of the past. It has been made obsolete by intensified specialisation – also reflected in the fact that academics will rarely venture outside the thin wedge of their expertise to engage in interdisciplinary research. And due to their quite distinct disciplinary divisions, universities are generally not attuned to conducting interdisciplinary research (Naznin et al., 2017). Specialisation has become the norm, but because of this, it is highly susceptible to redundancy. The capitalist machine will guarantee this as the search for greater profits through technology and automation cuts a swathe through the global workforce.

Studies reveal that up to 60 per cent of the jobs students are currently studying for will be obsolete within the next 20 to 30 years. Studies also indicate that in advanced economies approximately 20 to 40 per cent of jobs are at risk of being replaced by automation within the next 15 to 25 years. Considering this, one might surmise that there will always be a demand for good political leaders because of their scarcity. However, as artificial intelligence (AI) is realised, no job will be indispensable (see for example Erik Brynjolfsson and Andrew McAfee's book *The Second Machine Age*, 2016). This will have profound repercussions across the world and accentuates a paucity of integrity and/or foresight with respect to the bureaucracy of higher education and government policy per se. For instance, what policies are in place to ensure the 'relatively smooth' transition, or any transition, to an automated world? Furthermore, how does one reconcile an automated world with jobs for a potential ten billion plus people by the end of this century? Techno-unemployment on a massive scale is a likely norm.

The intense marketisation of education has reframed it in its likeness as a place of profit and the pursuit of material wealth. There is a definitive focus on research that is profitable to the detriment of all other research. In order to remain true to the cause, research proposals are written in a capitalist monotone and only sponsored if they represent a prestige-enhancing, moneymaking proposition. The criterion by which they are evaluated is simply 'profit and loss'. The exploitation of educators through the casualisation of the academic workforce in the quest to save costs and increase profitability is another case in point (see for example Freund, Kizimchuk, Zapasnik, Esteves, & Mewburn, 2018). The population of 'para-academics' continues to grow. Para-academics are the alienated and disenfranchised casualties of corporate academia. As Alex Wardrop and Deborah Withers (2014) opined: The demise of the university is apparent. Students listlessly file between classes as they struggle to achieve a means to an end, all the while burdened with the prospect of an intensely competitive capitalist marketplace hanging precariously over their heads. Moreover, 'The utterly dispensable para-academic watches this scene from the margins – subjected to the callous mediocrity of temporary contracts that offer absolutely nothing in terms of 'career development', or any kind of rung on the ruthless academic ladder' (p. 7). This is by no means a submission of surrender to the academy's neoliberal mindset, but it is no wonder that deep-seated feelings of betrayal arise.

Furedi (2013) pointed to university administrators as complicit in the infantilisation of their campuses: they have embraced 'narrow-minded and intolerant' 'illiberal practices' alienating students from the ideal of freedom. Furedi lamented that twenty-first century universities were constantly engaged in rule-making wherein the traditional freedom of free speech had been reduced to a submissive passivity. Our universities used to be places of pluralistic liberalism with the quest for knowledge, discourse, and learning, their goals. The freedom to finally express oneself epitomised the essence of university life. University life now mirrors the capitalistic qualities that neoliberalism emboldened and sanctified. Consumed by a dogma of corporate greed they have become dispirited and disengaged places. As philosopher Martha Nussbaum (2012) observed:

> democracies have great rational and imaginative powers. They are also prone to some serious flaws in reasoning, to parochialism, haste, sloppiness, selfishness, narrowness of the spirit. Education based mainly on profitability in the global market magnifies these deficiencies, producing a greedy obtuseness and a technically trained docility that threaten the very life of democracy itself, and that certainly impede the creation of the decent world culture.
>
> (p. 142)

A neoliberal system of thought has engendered a mad scramble post-haste not to be omitted from 'the competition'. As observed by Stephen J. Ball: 'neoliberalism gets into our minds and our souls, into the ways in which we think about what we do, and into our social relations with others' (p. 18). 'Neoliberalism', in testament to its significance to the economic growth imperative, was described by environmental scientist and philosopher, Alexios Monopolis (2012), this way: the predominant neoliberal economic development regime prioritises the values of the market – in contrast to those of social and ecological wellbeing – resulting in an uncompromising bias toward economic growth, the commoditisation of nature, the homogenisation of culture, the automation of most if not all activity, the material accumulation of wealth, human estrangement and manic individualism. It is most probably safe to say that Monopolis is not a fan of what we have come to understand as a neoliberal style of thinking and acting and governance.

So, what becomes of a society that marches docilely to the drum of a demented democracy, armed with university degrees that were mass-produced by those possessed of a neoliberalist mental model bent on the commodification of virtually everything? Have we unwittingly committed to the appropriation of our senses and our spirit, and thus the zombification of society?

Conclusions

Public policy cannot be separated from an institutional context. An institution is broadly defined herein as a cohort of social actors that prescribe to a set of shared mental models that ineludibly govern their interactions but not necessarily their thinking and reasoning. As such, institutionalisation might be considered a grand social experiment in which individuals might or might not reconcile their personally held mental model of reality with that of their institutions. Rebellious inclinations periodically arise. To be sure, institutionalisation means giving up certain freedoms, and the moral dilemma of what represents the more powerful incentive: to belong or to be true to one's beliefs. Especially if the freedom to choose, endorsed by the doctrine of democracy, is the overriding held belief? Mindful, also, that individuals do not exist in a democratic bubble; they have relationships that are sustained on a *quid pro quo* basis, extending beyond oneself to include family, friends, and social groups, which *ipso facto* blurs the boundaries between self-interest and altruism.

The aim to achieving a good policy outcome is to get from A (prediction) to B (policy) with the integrity and efficacy of the science and the policy intact. The Institutions Process Model (Figure 5.1) is a theoretical endeavour to explicate this process. It depicts three primary institutional mental models: the scientist, the layperson, and the policymaker; and two secondary institutional mental models: the politico-mediaist and the economist. All are obligated to, and bound by, the social ties of rational choice and public choice. All are in a constant struggle to impose their respective mental models of the issue of focus, in this instance, climate change, as the default mental model. Moreover, the power, and thus the influence an institution wheels, fluctuates over time.

And so, although von Mises (1998) described the freedom to choose with a sense of optimism, the freedom to choose ultimately rests on the balance of institutional power. Over time, one institution might exert more or less influence over the others, causing a good or bad imbalance. This accounts for the Age of Enlightenment that propagated tolerance, reason, and humility; and its nemesis, the Dark Ages, that propagated ignorance and prejudice. When an institution's capacity to act as intended (productively and ethically) is compromised by an extraordinary shift in power or persuasion, the universality of balance and stability is disrupted. When an institution becomes too powerful, or even all-powerful, the right to choose, for the majority at least, is lost. As economist and Nobel Prize recipient Friedrich Hayek (1960) proclaimed, it is the balance of power and cooperation between institutions that sustains the promise of individual liberty. Hayek, sometimes called the 'father of neoliberalism', might never have imagined, however, that the institution of individualism could be taken to such extremes. As a libertarian, Hayek's intent was presumably admirable – much less so has been the free-market fundamentalism that has ensued. And according to MacLean (2017), the same can be said of Buchanan and his public choice theory.

Irrespective of whether Buchanan and Co. are responsible for the current status quo, it is evident that our institutions have inspired a competitive dimension by providing the playing field and ground rules to 'casino capitalism'. A neoliberal governmentality aided by the power of public policy, as argued by Dardot and Laval (2013), has enabled the liberation of finance which in turn has 'engendered a 'creature' with a power that is simultaneously diffuse, global and uncontrollable' (pp. 158–159).

It is thus inevitable that the balance of power among institutions will not always remain in balance and members of an institution will often not agree with one another. The disparate mental models they hold of an object, scenario, or phenomenon are a source of discontent. Their thinking and acting and thus their institutional facts and narratives are potentially quite different. Subsequently, each institutional actor is free to frame the climate change object in accord with his or her worldview. Thus, climate change the incomplete and contentious idea, is fated to hold sway in some sectors of The Institutions Process Model until such time as climate change reveals its true self as a 'brute fact'. Indeed, at that moment, cooperation among institutions will presumably reach the object of necessity – or not?

Cooperation between institutions is premised on clear and concise communication. Ideally, a collective narrative in a common language founded on a set of 'shared mental models' is pivotal to the successful translation and transmission of scientific knowledge. Maintaining the integrity of the translation and transmission is an imperative. It is evident, however, that apart from the idiosyncratic vernaculars characteristic of each institution, the language of science as opposed to the language of politics might be construed as occupying two extremes of the authenticity spectrum. The objective versus the subjective as described by Parsons (1990). Opinion masquerading as fact is the scourge of contemporary politics and a prelude to the onset of another Dark Age in which all-powerful institutions endeavour to rule the minds, and thus the thinking and acting, of the publics.

Half-truths and hyperbole dominate political communications. Rhetoric in the sophist tradition is a common theatric in the theatre of politics. Differences of opinion are ubiquitous and deep-seated. Opinions tend to be experienced as truths, even though they often derive from one's own assumptions and politico-cultural background. In the dialogic process, basic assumptions and opinions are culturally specific and are often merged into 'institutional facts'. Furthermore, the exploitation of counterfactual theories of causation and compromise has the capacity to weaken the veracity of public policy. The prevalence of watered-down climate change policy is a case in point.

Relative to one's respective institution, one is either looking out or looking in. From these two perspectives, the scenario and behaviour of institutional actors looks markedly different: the social observer looking in judges the behaviour of others to be biased and irrational while the social actor looking out judges his or her behaviour to be unbiased and rational. Arguably, to be completely objective (unbiased and rational) is to be unhuman. Being unhuman is perhaps the providence of the scientist: 'Scientists speak in the name of science, not on their own behalf, and therefore the primary problem in describing those who speak mistakenly is to discern their motivation: their commitment to something other than science' (Jenkins, 2013, p. 70). The fallibility of being human might thus cloud the object of reason, but might also render one a 'rational fool' and 'social moron'. The same would apply to the layperson, the policymaker, the economist, and the politico-mediaist. Alternatively, perhaps removing the imposition of being 'only human' is the key to reflect more deeply on the complexity of climate change. *Prima facie*, the rationale is simple – to make plausible deductions from the truth in the premises or the validity of the cues.

Although public choice is relative to all five institutions in the Model, it is reasonable to suggest that public expectations are less forgiving with regard to the policymaker and as such, the politico-mediaist. Their sincerity, in the sophist tradition, is not beyond reproach. Mindful that science and politics are apt to pursue different paths. They speak a different language. The endeavour of the scientist – and thus the institution of scientific reality – is to cast out the rhetorical form. To strike down the 'language of bullshit' as described by Frankfurt (2005). Whereas the endeavour of the politico-mediaist (and thus the politico-media complex) is to master the art of sophistry and use it as the great masters would use a brush to create their respective image of the world. This mental image of the world, that is more or less institutionalised and a compilation of the mental models held, lies at the crossroads of the preservation of scientific knowledge and one's ideological bias and thus, aids or impedes the design and implementation of deliberative climate change policy.

In defence of the adopted political system and its caretaker the politico-elite, we might cite Thomas Hobbes' argument, from his classic work, *Leviathan* ([1651] 1981), that in the absence of sovereign rule people are not so altruistic or caring, that we are motivated by desire and aversion with self-preservation our priority. We might even revert to savagery to attain it. Hobbes maintained that it is irrational to think that human nature would operate otherwise. Political actors, along with the rest of the cast, do indeed act rationally by pursuing their own self-interest – beneath the façade of roleplaying we are all 'egoistic, rational, utility maximisers' at heart. They pursue their own self-interest in order to maintain order and quell the unrest of others pursuing their self-interest to the detriment of others doing the same.

All that said, one cannot just brazenly dismiss altruism – behaviour that costs the actor and benefits the recipient. According to Kurzban et al. (2015), altruism is a commonplace prosocial behaviour. Arbitrary acts of collective altruism thus confound rational choice theory. Whether the act of altruism is motivated by sympathy, compassion, communitarianism, cosmopolitanism, feelings of guilt, or the need to feel better about oneself, or some innate urge to do the right thing, is of no consequence to the outcome.

It is also evident that despite the pervasive institutionalisation of the publics, to a particular political persuasion, more effortful thinking and reasoning and the acquisition of knowledge has the potential to further diminish the intended effect of what might be the duplicitous politico-media framing of an issue. Besides, given the publics' general distrust of politico-elites and government institutions, it is doubtful that the well-educated publics are so credulous to believe, without further inquiry, the constant flow of dubious political claims and counter claims that ply the politico-media airwaves. Mindful, also, that escape to the virtual world is only a click away.

As Mark Deuze (2013) observed: 'All this intense and immersive media use can be seen as turning us into helpless addicts, slaves to machines – *zombies*' (p. 1). Thus, for those who have adopted the collective mental model of the 'viral media zombie' and claim the virtual world as their common abode, the framing of a political communication from the real world might be lost in translation, fail to reach its target, or be simply ignored. Anaesthetised by the constant flow of political obfuscation, claims and counter claims, character assassinations, the infantilisation of democracy, policy paralysis, and a failure of our institutions to do as they were intended. The publics donned the headphones and escaped to the peace and tranquillity of the virtual world. The glitch is, of course, that the virtual world is also a playground for propaganda, bias, untruths, and bad science.

In reality, although the virtual world has its appeal, the publics have in no way freed themselves from the clutches of public and private institutions. As long as we remain more or less institutionalised we are captured and controlled, to some extent, by institutional power. Power corrupts, and public and private institutions are highly susceptible to corruption as history shows. Anticorruption 'watchdogs' have been established to combat this with only marginal success. Michael Johnston (2014) described corruption succinctly as 'tenacious and sustained by powerful incentives'. The push for increased discretionary powers by political ideologues accentuates the corruptive influence of power. Research has shown that corruption, through illicit inducements such as bribes to public officials by vested interests, amongst numerable other things, adversely impacts on the integrity of institutions charged with implementing environmental protection, and has a stifling effect on the ratification of environmental protocols. It is also a major contributor to growing inequality, and instrumental in maintaining the status quo by keeping the poor and marginalised, poor and marginalised.

Governments have a 'duty of care' to protect their citizens from dangerous climate change. Consequently, if scientific reality pointed to climate change having a harmful impact on human welfare, and governments had the capacity to avert such harm, they are morally and legally bound to do so. Based on the premise that framing the climate change problem is *prima facie* a reflection of the mental model held, I surmise that failing to act on climate change can be attributed to a deficient mental model of the climate change problem – bearing in mind that a mental model is a blueprint for thinking and acting. Indeed, failing to act in advance of the threat of dangerous climate change is not just a failure of our institutions, but is no less than a crime against humanity.

Notes

1 The 2016 Global Green Economy Index (GGEI) provides a comprehensive assessment of how 80 countries fair in the global green economy. Countries are measured by expert practitioners who rank their performance in terms of perception and performance in environmental areas such as climate change, air quality, water, forests, and agriculture: available at http://dualcitizeninc.com/global-green-economy-index/.

2 I note, however, that Oxfam's methodology is not normally open to peer review; hence, their calculations may be questionable when measured against more rigorous empirical standards.

6 The Mental Model Process

Hunting magic is one theory underlying Upper Palaeolithic cave art (≈40,000–28,000 BP) discovered in, among other places, Lascaux in France and Altamira in Spain. To simulate a successful hunt, representations of the hunted animals were depicted on cave walls. Hunting magic theory was perhaps a nascent phase of mental-modelling. Moreover, the survival of ancient hunter-gatherers was predicated on creating a detailed mental model of their surrounding territory – knowing the life cycles of plants and animals, and their habitats, was essential information for their ongoing subsistence (Harari, 2011). The more contemporary mental models we carry in our head are symptomatic of our goals, opinions, values, beliefs, cultural background, and current physical disposition and 'place' (the physical environment in which one is located). And seen through such ideological eyes, an object, scenario, or phenomenon might be reimagined, if it is altogether displeasing or even abhorrent to the observer. Or perhaps it is too complex to model, and the observer might simply reimagine it in accord with his or her shared cultural codebook (explained in more detail below), arbitrarily dismissing any falsifying models or even abandoning the search altogether.

For example, Alan Greenspan's belief in a self-regulating market crashed with the 2007–2008 GFC. As Meena Thuraisingham (2013) affirmed: 'Our singular commitment to a set of values or beliefs can blindside us in ways that can have catastrophic consequences' (p. 21). In Greenspan's case, the shock of the GFC resonated strongly enough for him to reflect on the validity of the mental model he held of how the global financial market actually works, and consequently abandon his pre-GFC mental model as deficient – albeit belatedly. His conviction of a free-market institution was not dissimilar to beachside residents rebuilding for the second or third time in coastal zones exposed to the ferocity of severe tropical storms: Katrina, Harvey, and Irma, for example. They hold to a mental model that rejects intensifying storminess, sea level rise, and storm surge as a genuine threat to their way of life, and often ignore directives to evacuate on the basis that it 'has happened before and we survived'; nothing has changed. In many cases local government officials are unwilling to even talk about climate change adaptation policy because of concerns about their city council's legal liability. A planned retreat from the coastline, for example, would adversely impact on the chosen way of life and possible livelihood of beachside dwellers and the value of their properties, in addition to causing stress and anxiety to those responsible for enforcing it. So, how do these models that accept and/or reject certain information that provides the impetus to orchestrate our thinking and acting – shall we stay or shall we go, for example – actually form?

The Model

The Mental Model Process (Figure 6.1) extends from the Institutions Process Model (Figure 5.1), and was derived from the works of Craik (1943), Kaku (2014), and Johnson-Laird (1983, 2006, 2013) on mental-modelling and cognitive science. The Model moves through a series of phases and filters not necessarily in the order depicted. I describe the image search as automated and constant as the brain constructs and deconstructs mental models of the world and then simulates them into the future. Sleep would naturally suggest a state of unconsciousness whereby the image search is disengaged. However, as Patricia A. Alexander (2016) explained, it matters not whether an individual is awake or asleep, or indeed conscious of their thoughts and perceptions, mental processing is constantly active.

Research has revealed that during 'lucid dreaming' one is aware to some degree and capable of directing the content of a dream. John Cacioppo and Laura Freberg (2014) defined dreaming as: 'A mental state that usually occurs during sleep that features visual imagery' (p. 209). According to Kaku (2014), dreams are simply an alternative state of mind of simulating the future, 'one in which the laws of nature and social interactions are temporarily suspended' (p. 174). More precisely, as humans drift into sleep the brain stem engages another system called the cholinergic system. This system emits certain chemicals that induce a dream state. Cholinergic neurons set off erratic electrical pulses that stimulate the visual cortex to create dreams. These erratic pulses, however, cause cells in the visual cortex to resonate irregularly at hundreds of times per second, which might account for the often-incoherent nature of dreams.

Importantly, the cholinergic system also emits chemicals that decouple the parts of the brain involved in reasoning and logic (Kaku, 2014). This means that although dreams provide the domain to suspend such things as gravity, and have been implicated in predicting the future, their rational interpretation remains speculative. According to psychoanalyst and neuropsychologist Mark Solms (2014), in reality, dreams are purely subjective phenomena and most of the literature explaining them is fragmented and disjointed. I surmise that dreaming might be mistaken for mental-modelling. The two might commingle on occasions which confounds the innate desire to decipher their meaning. There is a fundamental difference however: the latter represents consciousness, whereas the former represents only partial consciousness at most. And the key to accurate mental-modelling is a high level of consciousness. As such, I suspect that Leonardo was not dreaming of flying machines, but mental-modelling them.

The arrows in the Model shown in Figure 6.1 represent a possible account of 'process flow'; however, back and forth, multidirectional, and simultaneous

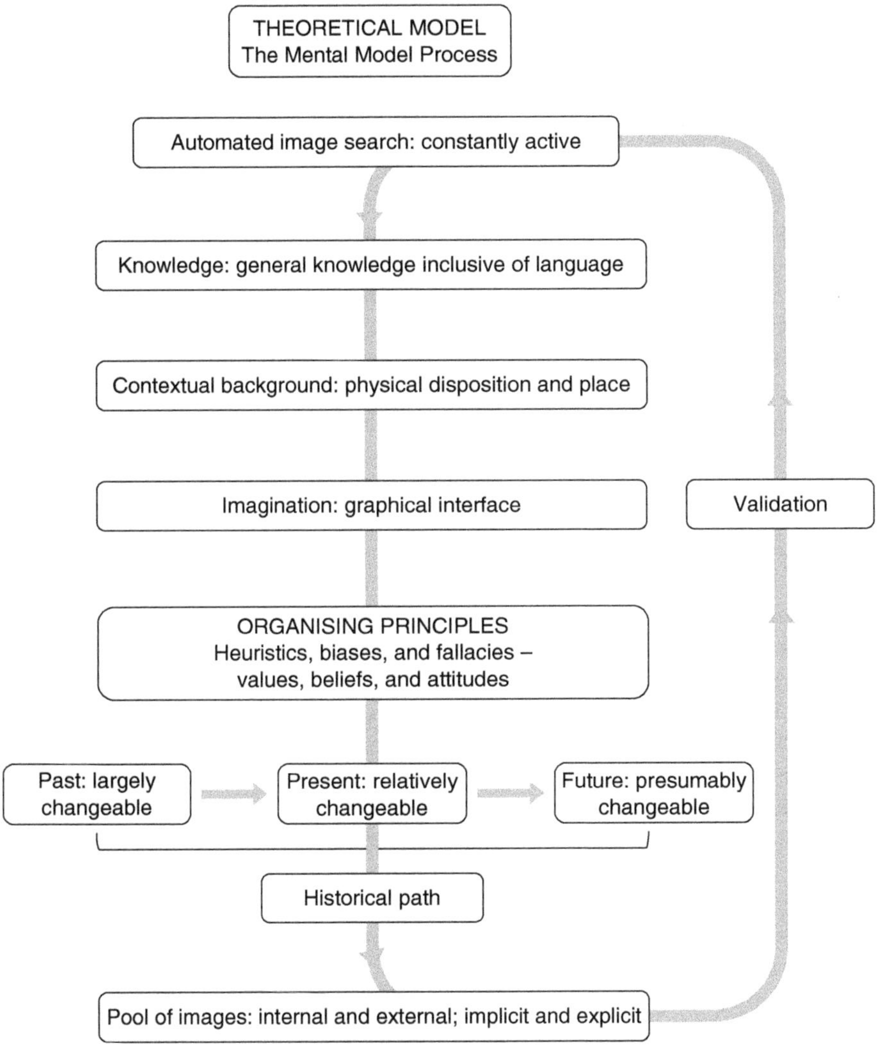

Figure 6.1 The Mental Model Process.

projections are probably just as likely. In the sense that 'Our imaginations project a future that we remember and thus becomes our past' (Lipari, 2014, p. 134). Indeed, a consensus as to how the mind works, the underlying processes of thinking and reasoning, is yet to be achieved. As Kaku (2014) concluded: in all of nature the two greatest mysteries are the mind and the universe.

The 'mind is understood as the firing of neutrons in an effort to create an internal model [mental model] of experience and the world' (Hall, Curtin, & Rutherford, 2014, p. 20). In contrast to the brain which is described by Hall et al.

(2014) as a 'mechanism of learning that computes and calculates data to an internal model that then has to interpret and adapt based on past and future updating (learning)' (p. 20). The pursuit of learning, however, is contingent on perfecting these internal models stored in the brain. Suffice to say, there is often a strong proclivity to retain the old model. A 'learning' definition of the brain, as cited in Hall et al. (2014), is as follows:

> learning means updating an internal model generated and maintained by the brain about the world. The brain is distant from the world. All it has are sensory inputs that are ambiguous, noisy and the brain has to make sense of them. Any given particular input obtained needs an image encoded by neurons in the brain and then the brain has to interpret something causing it in the real world. In order to do that it has to have a model. That model is maintained and updated by the brain in light of prediction arrows and that updating is learning – changing of the parameters of the mathematical model by the physiological architecture of the brain.
>
> (pp. 19–20)

Updating one's mental model, in the process of learning, sounds simple enough. However, a long-held mental model fashioned and framed over a lifetime might not be so amenable to change. Especially if it requires a significant modification to one's 'orderly and usual' way of life and the abandonment of old patterns of thinking and acting. For instance, there is a profound probability that an individual's mental model of how the world works is not how the world works. As Ronnie Janoff-Bulman (1992) surmised, in the instance of renewal and revelation, 'The very assumptions that had provided psychological coherence and stability in a complex world are the very assumptions that are shattered' (p. 64). Fear and anxiety might subsequently ensue on discovering that what we thought we knew about the world was only a semblance of its truth. Updating or modernising an anachronistic and thus deficient mental model might then entail a traumatic event – a challenge to one's intrinsic desire for the world to be predictable.

Predictable to the extent, that in addition to not only not knowing how the world works, the world we think we live in is not the world we live in. That is, although the language of mathematics, as a pretext to discovering and communicating reality, is now conspicuously confounded in quantum theory, which renders numerous presumably established theories incompatible by way of the scientific method itself. Indeed, the 'quantum model of nature encompasses principles that contradict not only our everyday experience but our intuitive concept of reality' (Hawking & Mlodinow, 2011, p. 74). Quantum theory deconstructs science as we understand it to be – objective knowledge. The world we remember and know is almost gone. But this is another story.

Despite our penchant to cling to past pretexts, the brain is a 'prediction machine'. One of its core functions is to make action-related predictions about the environment. Andy Clark (2013), in his endeavours to unify the brain's functions, has called this 'action-oriented predictive processing', which allows humans to interact with a changing environment [a changing climate, for example] in an adaptive way. This constant of human adaptation is often referred to as the 'environment of evolutionary adaptedness'. The environment of evolutionary adaptedness is not confined to a particular space and time but is contingent on the particularities of the adaptation required to fulfil the fundamental need for survival. Adaptation through evolution, then, better designs humans to survive impending perils. Adaptations are adopted, exapted, or discarded to enhance fitness. This adaptive process takes place in geologic time; as such, there is a time-lag wherein the adaptive traits are periodically standardised in the mental models that humans hold as the mind endeavours to catch up to contemporary adaptive problems that periodically transpire.

Mental-modelling 101

Our brains are exceptional mental model builders (Hawking & Mlodinow, 2011). For instance, if a person is fitted with glasses that turn the images he or she sees upside down, the brain will, after a time, modify its mental model so that the image appears the right way up. When the glasses are removed, the world appears upside down for a time, the brain then adapts by correcting its mental model of the world so it is once again the right way up. Viewing the world the right way up is a good place to start, but how does one construct an understanding of that worldview in order to navigate one's world journey? Moreover, how does one's worldview affect one's thinking and acting?

At its most basic level, mental-modelling can be viewed as the science of reasoning. People commonly reason in what appear to be totally illogical ways. The reasoning that underlies the rationale that underlies a person's behaviour is often far from rational. According to Paul Feyerabend (1981), the conclusion of intelligent and thoughtful observers of human behaviour is that it is naïve to believe that the simple-minded rules plied by methodologists could be capable of accounting for the complexity of the human condition. Indeed, a consensus as to the underlying processes of advanced thinking has yet to be reached. For instance, what should be a clinical science, of separating knowledge from belief, is not so clinical, as explained by David Krech and Richard S. Crutchfield (1948):

> There are no impartial 'facts'. Data do not have a logic of their own that results in the same perceptions and cognitions for all people. Data are perceived and interpreted in terms of the individual perceiver's own needs, own connotations, own personality, own previously formed cognitive patterns.
>
> (p. 94)

The mind is thus never truly objective and open to embracing incoming data. Incoming data is intercepted by our sensory organs and interpreted by our brains in accord with our own unique set of organising principles. The mental model formed through this interpretation is a customised version of reality. There is no model-independent test of how closely this customised version matches reality (Hawking & Mlodinow, 2011). Presumably, this is contingent on the user's level of consciousness, reasoning proficiency, and cognitive exertion.

Mental-modelling is a reasoning process that prescribes a particular way of thinking and acting. It is a means to an end and the fundamental process of giving meaning to premises. Premises are hypotheses, propositions, or evidence derived from external and introspective observations. Premise models are personally held integrations of the information comprised in premises which consist of inferred beliefs, both true and false. This means that rational and logical mental-modelling requires that beliefs are true in order to reach a valid deduction. A valid deduction – a plausible conclusion to the mental-modelling process – takes the form of a mental representation. A mental representation provides the blueprint for thinking and acting in response to a given premise(s). Notably however, by most accounts, a clear distinction between a mental model and a mental representation (or premise model or an isotropic model) is not apparent (Johnson-Laird, 2013; Matthen, 2014; Orlandi, 2014). In light of this, I shall treat them similarly.

A mental representation is not easily defined. Nico Orlandi (2014) described mental representations as 'unobservable entities and their presence and characteristics can only be inferred through the behaviour that they supposedly bring about and explain' (p. 7). They differ from conventional representations such as 'maps, paintings, photographs, signs, and stories', but share some of the same qualities. Mental representations might be associated with conventional maps or blueprints on the basis that they can be associated with a particular pattern of thinking and acting. Extending this, Orlandi defined mental representations as internal intermediaries that inform about some state of the world: they are derivatively abstract functional notions, as they affect an individual's performance, that help explain a person's behaviour. It is evident that Orlandi could just as well be referring to the mental models described herein. I would however question her theory that 'vision is not a cognitive process' (see for example Mole & Zhao, 2015). I suspect that seeing is a form of problem-solving, and thus reasoning, as discussed earlier. There is an inferential link between what we see and what we know.

A mental representation might be construed as a mental image. Mental images are constructions of the imagination which means they are subject to comprehension, interpretation, and conceptualisation. There is no clear consensus as to the mechanisms that constitute and organise mental images (see for example Kosslyn, 1994, 2005; Pylyshyn, 1981, 2002; Richardson, 2013). They are multidimensional metaphors, reflections, ideas, impressions, and predictions. Neuropsychologist and neurophysicist, Karl Pribram (2011), proposed a hypothesis of mental images with his 'holonomic brain theory'. According to Pribram, the brain stores an image in a holographic format and each part of a three-dimensional hologram comprises an image of the whole. This means that the brain has self-organising properties that can reconstruct the same image from multiple points within the brain. Incoming information is merged with stored images in a holographic-like matching process. One must then presume that in order for an individual to function, even to the minimum requirement of the everyday, some degree of updating is naturally taking place. Whether we are, or have the capacity to be, consciously aware of this is another story.

A mental representation (or mental model) might be the source of numerous mental images. Mohan Matthen (2014) explained that images derive from an underlying mental representation – which he referred to as an 'isotropic model' – that generates various perspectives enabling the perceiver to zoom in and out and modify some aspects. For example, a person is able to generate an image of how their driveway looks from the roof of his or her house without ever viewing it from that perspective. The driveway is subsequently created from a compilation of the perspectives generated, irrespective of whether the perceiver has actually experienced it from all those angles. This suggests that perception works in tandem with imagination: 'What is there implies what is not there' (O'Connor, 2015, p. 15).

Imagination fills the gaps left by perception. Suffice to say, climate change is substantially more complex than the driveway of a house. Nonetheless, the perspectival view of the climate change object, that might never have been physically experienced, is progressively built into its mental representation. For this to happen some measure of an underlying mental representation – in this instance, some perspective of climate change – must already exist, whether construed from popular opinion, the politico-media, education, imagination, or some other form of past experience.

Implicit and explicit codes, accompanied by a corresponding set of images, are involved in mental-modelling. A mental model, then, might be described as a 'how to do manual' – a 'handbook to life' in accord with Freud's critique discussed earlier – for thinking and acting which is progressively enhanced by the compilation of relevant images. But a manual is only meaningful and of use if we can accurately decipher the instructions, updates are made regularly, and the images are representative of reality. Mental-modelling commences with an image search prompted by an innate principle or impulse to formulate a rationale in response to external and/or introspective observations. This perceptual process comes naturally. It involves exploration and collation. The goal is to reach a plausible conclusion or scenario from a pool of possible alternative mental models as their corresponding images are collated. From an initial set of implicit models a more explicit model is drawn. Through more effortful reasoning, this explicit model is tested for validity against other models based on similar or the same premises that might falsify it. Validation is achieved when a falsifying model cannot be found. The validity in the Model, however, is relative to the cognitive effort expended during the search for a falsifying model

Knowledge is central to more effortful reasoning; it comprises the medium by which all our cognitive processes operate (Varga & Bauer, 2014). Knowledge underlies thought and allows us to make inferences and gives rise to language. The boundaries of human conceptualisation, however, stretch well beyond the knowledge that is directly acquired through learning. Indeed, there is a human capacity to self-generate new knowledge by identifying relational similarities abstracted from episodic experiences; to decontextualise this knowledge and store it as integrated representations (mental models) in semantic memory. Semantic memory is that component of memory responsible for acquiring, representing, and processing information (Saumier & Chertkow, 2002).

This has implications for the human mind as an autonomous computational machine, and thus Johnson-Laird's (2013) more recent explanatory theory of a quasi-mechanistic approach to human thinking and reasoning. However, the idea that we can self-generate new knowledge suggests a capacity more advanced than a computational machine. A point to note is that Johnson-Laird (1983) originally conjectured that reasoning was an imaginative skill. This makes more sense to me. I would submit that it would be unwise to discount the imagination as instrumental to the reasoning process. Let the imagination run wild I say! Until, at least, it is constrained by some insurmountable reality.

I surmise that the computer analogy – a mechanistic approach – is rather limiting and unimaginative. The inference of the mind remote from the world – alienated from sensory receptors and effectively insulated from the external environment – is positively unappealing. What is more, the mind/brain cocooned within the head in a primal state amounts to nothing more than a zombie state, one would suspect. It is much easier to imagine that our state of mind and behaviour have direct links to the immediate environment (discussed in more detail later). Our independence from the environment, both mentally and physically, makes no sense as we think and act very differently when exposed to acts of violence as opposed to acts of kindness – in anticipation of the act, at the time of the act, and in the aftermath. Mindful that the mental models we hold are predictive of our thinking and acting and as such, the emotions we might experience. In this sense, a changing environment would also presumably lead, eventually anyway, to a change in the mental model held in order to prescribe the appropriate thinking and acting and emotions for that changing environment.

So, although we might appear as products of socialisation and culture and therefore of our immediate environments – but not necessarily passive recipients – our thoughts and actions and their corresponding emotions are based on predictions of how to think and act when exposed to fluctuations in those environments. An error occurs when the mental model we use is obsolete and the corresponding thinking and acting and emotional response is inappropriate for the moment and also the predicted environment. The key, then, is to endeavour to ensure that the mental model held is not only a true reflection of one's immediate environment, but also predictive of any change that will transpire – an earth-shattering climatic change for example.

Nicola Tesla: a mental-modelling supremo

I would add to the list of supreme mental-modellers, among those such as Leonardo and Galileo, the enigmatic innovator and eccentric Nicola Tesla. More than 100 years ago Tesla invented alternating current (AC), the polyphase alternating current system, thus creating the mass-produced power supply we have today. He proposed a global system of electrical energy supply that was clean and potentially inexhaustible, but was constrained by the laws of nature and persons he described as 'ignorant, unimaginative, and consumed by self-interest' (Carlson, 2005; O'Neill, [1944] 2007; Rybak, 1999). From the invention of the particle beam to radar, the electric car, robotics, and remote-controlled drones, Tesla was able to mental-model the solution to a problem and engineer designs with such clarity of mind that he reportedly visualised the individual components in three dimensions, ran simulations and checked for wear and tear.

Tesla ostensibly reproduced mental models of his inventions in much the same way as described by Pribram (2011) and his holonomic brain theory – as a three-dimensional holographic image. One of his greatest exploits was to transmit electrical power over considerable distances without wires or cables, something that has baffled scientists to this day. Tesla was convinced that the Earth comprised 'fluid electrical charges' that when interrupted by a series of electrical discharges at repeated intervals would produce an abundant power supply by generating immense low-frequency electrical waves. His grand vision was to free humankind from the 'necessity of mining, pumping, transporting and burning fuel, and do away with innumerable causes of sinful waste' (Tesla, 1905, p. 7).

Suffice to say, Tesla was seeking a source of inexhaustible clean energy and was ardently opposed to centralised coal-fired power stations that spew CO_2 into the air that humans breathe. John J. O'Neill ([1944] 2007) wrote in his biography of Tesla:

> The panorama of human evolution is illumined by sudden bursts of dazzling brilliance in intellectual accomplishments that throw their beams far ahead to give us a glimpse of the distant future, that we may more correctly guide our wavering steps today.
>
> (p. 3)

Given Tesla's superior aptitude for advanced mental-modelling, it follows that he also had a mind for deep reflection – System 2 reasoning.

System 1 and System 2 reasoning

The intuitive mind predates and pre-empts the reflective mind. Jonathan Evans is perhaps the most cited and prolific publisher of research on 'dual-process theory', referred to herein as System 1 and System 2 reasoning. Elsewhere, for example, the terms 'two minds theory' or 'Type 1 and Type 2 processing' might be used. Moreover, Evans (2008) has reviewed distinctions among cognition styles that approximate System 1 and System 2 reasoning. For instance, 'automatic detection and controlled search' (Schneider & Shiffrin, 1977); 'holistic versus analytic cognition' (Nisbett, Peng, Choi, & Norenzayan, 2001); 'impulsive system versus the reflective system' (Strack & Deutsch, 2004); and System 1 and System 2 processing (Stanovich, 1999, 2004). I have adopted the latter herein as it is the one most apt for my story.

An early empirical investigation into dual-process theory per se – alternate pathways of cognitive processing – was conducted by social psychologists Richard E. Petty and John T. Cacioppo (1986). Much in the frame of System 1/System 2 reasoning, but with a focus on persuasion, their Elaboration Likelihood Model (ELM) defined two pathways of cognitive processing: the central route and the peripheral route. The former required considerably more cognitive elaboration than the latter, resulting in a change in attitude and/or behaviour. The latter takes a more superficial pathway that engages in simple heuristic cognitive processing, resulting in predominantly transitory attitudinal and behavioural changes. Moreover, with respect to the latter, the elements of likeability and trust (the use of celebrities to endorse consumer products, for example), in terms of the source of information, might override the motivation to think more deeply about the validity of the information. Presumably, in order to validate a truthful mental model of the object, scenario, or phenomenon of focus, one should take the central route. It might be longer but should prove safer. Determining the validity of the outcome as a consequence of the route taken, however, is not so clear-cut. Taking the central route might yield a relatively objective outcome, or an equally biased one (Cacioppo, Cacioppo, & Petty, 2017). Importantly, Petty and Cacioppo's ELM has received substantial empirical advocacy by proving relevant in clinical practice.

Constructive memory informs the mental models held. David F. Sherry and Daniel L. Schacter (1987) argued the evolutionary hypothesis of multiple memory systems – which they described as System 1 and System 2 – on the basis of 'functional incompatibility'. They proposed that multiple systems adapted to perform what seemed to be similar functions but actually differed in important ways. The evolution of multiple systems was a consequence of environmental demands that required specialisation. Thus, a functional incompatibility occurred that necessitated the evolution of multiple memory systems: simply stated, the intuitive versus the reflective.

Cognitive psychologist Arthur S. Reber (1989) described this evolutionary distinction more succinctly as 'primitive' (in reference to System 1) and 'sophisticated' (in reference to System 2): 'Consciousness [as it relates to the sophisticated mind], evolutionary speaking, is a late arrival on the mental scene' (p. 230). Evans (2014b) explained that 'there are older, animal-like forms of cognition which combine with more recent and distinctively human kinds' (p. 131). The old mind is an ancient learning system that is shared with other animals; whereas the new mind, which is predisposed to science, engineering and literature, evolved in modern human beings.

In the 1960s Kagan, Rosman, Day, Albert, and Phillips (1964) conducted a series of eight reflection/impulsivity studies on school children (grades 1–4) (see also Kagan, 1965), inspired by the question: 'Why are some children more reflective than others?' (p. 33).[1] They focussed their research on three possibilities: 'constitutional dispositions, degree of involvement in tasks, and expectations of failure' (Kagan et al., 1964, p. 33). They found that children who did not reflect on the 'differential validity' of the set of solution possibilities were prone to settle on the first idea (or by extension, mental model for the purpose herein) that occurred to them, which was more likely to result in failure. Whereas children who chose to reflect were more likely to succeed. Moreover, those children caught in a 'cul-de-sac' in a problem-solving sequence (with no obvious path to further progress), upon recognising that they had not succeeded in solving the problem, became more anxious than they were at the beginning of the sequence. Although Kagan et al. (1964) argued that these findings had 'generality and stability', they did not elaborate on the implications. For example, it remained unclear how these findings 'carried over' from children to adults, if at all.

Research by Langer, Blank, and Chanowitz (1978) found that unless a communication was explicitly structured to elicit an effortful response, ignorance of relevant information would occur. This means that relevant information might be bypassed in favour of information that was consistent with a past experience. Langer et al. referred to this as 'mindlessness', but declined to conclude that it was the rule rather than the exception. For now, we can think of mindlessness as invariant or habitual behaviour largely devoid of conscious awareness. Mindfulness, its opposite, would prompt us to 'reflect' before making a decision that has an element of risk.

In the absence of mindfulness, humans are willing to risk the future outcome of a decision by avoiding a true evaluation of its consequences. Mats Björkman (1984) maintained that 'man is captured in a narrow time-space sphere' around the 'here and now' the consequence of which is that 'risk-willingness increases the longer ahead the outcome of a decision is likely to occur' (p. 31). The less involved individuals are, and the less knowledge they have, the more likely they are to make decisions that involve future risk. The decision to abandon pledged CO_2 emissions reduction targets fits neatly into this equation. It is difficult to comprehend the veracity of a decision taken now if its consequences lie far off in the future. In short, the human mind is programmed by the past, captured by the present, while endeavouring to predict the future. And human survival hinges on the accuracy of predictions about the future. Paradoxically, to be sure, even though we are captured by the 'here and now', if we were more mindful of it, we would be more adept at planning for the future.

From a technical perspective, as Kaku (2014) put it:

> the link between the dorsolateral prefrontal cortex and the hippocampus lights up when a person is engaged in planning for the future *and* remembering the past. In some sense, the brain is trying to 'recall the future,' drawing upon memories of the past in order to determine how something will evolve into the future.
>
> (p. 113)

Thinking and reasoning about the future consequences of today's decisions, particularly when the object, scenario, or phenomenon is imprecise or complex, requires a conscious effort to engage in hypothetical reasoning that is common to System 2. Kaku (2014) believed 'that human consciousness involves the ability to create a model of the world and then simulate the model into the future, in order to attain a goal' (p. 138). An individual's level of consciousness is, therefore, associated with his or her capacity and proclivity to engage and remain in System 2. By extension, a 'true visionary' would possess a high level of consciousness. Einstein, for instance, was known to devote up to a decade to a thought experiment in order to realise a plausible conclusion (his theory of special relativity, for example). He is quoted as saying 'I have no special talents … I am only passionately curious' (as cited in Calaprice, 2011, p. 20). According to Kaku (2014), however, 'Einstein's genius was probably his extraordinary ability to simulate the future through thought experiments, creating new physical principles via pictures' (p. 133). One can liken this to Tesla's extraordinary ability to precisely mental model the objects and scenarios that he would later create.

Thinking and reasoning are clearly not a simple outcome of cause and effect. A myriad of implicit and explicit controlling factors (or organising principles) operating simultaneously and at multiple levels are complicit in the outcome of the reasoning process. A pattern of thinking will probably emerge, but this does not explain an individual's disposition to jump to System 2 reasoning as opposed to his or her disposition to remain in System 1, which ultimately determines the degree of validity in a deduction. The assumption being that an invalid mental model, pursuant to an implausible conclusion, is more likely to result from System 1 reasoning. What instantiates or triggers the disposition to jump to System 2 is decidedly more complex. There is a natural predilection to immediately engage System 1 and remain there as the least effortful enterprise. This is consistent with cognitive economy and the cognitive miser model discussed in Chapter 1.

The concept of System 1/System 2 reasoning might, in part, explain the claim by scientists that the publics' grasp of scientific knowledge was inadequate, and that a general reluctance to 'find out' rendered the language of science broadly incomprehensible to the layperson or non-scientist. Similarly, with respect to validating political communications, fact checking every claim by every politician would require considerable effort. The uncertainty in the science and the obfuscation in political speak probably provide enough of an alibi not to pursue more effortful thinking and reasoning. Uncertainty and obfuscation do not inspire one to act, and likely do the opposite. Indeed, under this pretext, acting in response to climate science is probably counterintuitive in the everyday.

With respect to rational choice and public choice – prevalent in decision-making and, I submit, that there is no discernible difference between the two – they are often mistaken for the general proclivity to remain in a System 1 mind mode. For example, System 2 reasoning has the capacity to override cognitive biases such as 'self-interest', which is commonly demonised as the universal scapegoat for humanity's moral and ethical failures. It follows, then, that thinking and reasoning in a mode wholly governed by self-interest can only be regarded as shallow thinking (as in System 1). In summary, as noted by G. Pennycook, Cheyne, Seli, Koehler, and Fugelsang (2012), 'An analytic cognitive style denotes a propensity to set aside highly salient intuitions when engaging in problem solving' (p. 335). 'Thinking style', then, describes a person's preference for either System 1 or System 2 that he or she relies on to negotiate the everyday. The optimal application of which is contingent on the architecture, or organising principles, of the mind.

The cultural codebook

The worldview that individuals hold is a function of how they organise their thoughts. Information is processed into thoughts. The human brain, however, is not so adept at processing the vast quantity of sensory stimuli available to the sensory organs. It is estimated that only 2 to 5 per cent of incoming sensory data is processed and the rest is filtered out (McDowell, 2015). Knowledge, whether a posteriori or a priori, inclusive of knowledge of language, is requisite to comprehending incoming sensory data and formulating an initial set of quasi-compliant implicit images. Mental images are that part attributable to the imagination. The imagination is a graphical interface and a bridging mechanism. In accord with one's cultural codebook, it translates the messages received from the senses (sensory organs) and fills in any gaps left by perception.

One's cultural codebook provides the organising principles for mental model construction. I refer to organising principles as the filtering and classifying processes of the mind. They are a composite of heuristics, biases, and fallacies, and values, beliefs, and attitudes, extending to other interactive and interconnected functional states such as perceptions, memories, intentions, desires, and fears, that institute daily social realities. Importantly, like good architecture in the functioning of a building, good cognitive architecture is central to the functional organisation of a coherent set of organising principles. An individual's organising principles are intimately interwoven into that person's perception of reality. Organising principles give meaning to sensations. Sensations are constantly exposed to heuristics. Heuristics are more plainly described as 'rules of thumb'.

Kahneman and Tversky (1973) are credited with the most influential studies of heuristic processes. People 'rely on a limited number of heuristics which sometimes yield reasonable judgments and sometimes lead to severe and systematic errors' (Kahneman & Tversky, 1973, p. 237). The outcome of heuristics can be described as something attune to validation (or an illusion of validity), which leads to an individual's hypothesis of reality. The glitch is that although heuristics are analogous to a fast computer processor, they are shortcuts that have a propensity to overlook important information. Simon (1956) described this as satisficing. A satisfactory conclusion might be reached, insofar as the degree of validation achieved is only equal to the aspiration to do so.

Some organising principles are invariant; however, they remain predisposed to the influence of culture and are mutable to some degree. In reality, the extent to which culture infiltrates our senses and orchestrates our thinking and acting is perhaps indefinable. Cultural researcher, anthropologist and scientist, Edward T. Hall ([1959] 1980), declared that 'The idea that man as a cultural being is bound by hidden rules and is not master of his fate may come as a shock to some' (pp. 119–120). Culture, according to Lisbeth Lipari (2014), 'is more than values, ritual, custom, belief. It is more than art and architecture, clothing or music. Culture is a living being and its habitat is the body' (2014, p. 30). That said, some people could be considered cultural chameleons. Their malleable minds allow them to blend into almost any culture.

The organising principles of the mind are implicated in both System 1 reasoning (shallow and intuitive) and System 2 reasoning (deep and reflective). It is also evident that different organising principles are engaged in System 1 and System 2, as they are used, and designed, for different purposes. For example, two-minds theory suggests that instinctive responses, intuitions, belief biases, linguistic cuing and learned habits are quickly accessed and are readily available in System 1. They automatically chime in and prompt us to 'jump to conclusions' or 'jump out of the way'.

In his endeavours to explicate System 1 and System 2 reasoning, Evans (2010) suggested that humans have two distinct minds within their brains: 'one intuitive and the other reflective'. The intuitive mind is primitive and shares many of the adaptive types of behaviour of our ancestors. The reflective mind, by contrast, is more recently evolved and enables us to think in abstract and hypothetical ways about the world, and to calculate the future consequences of our actions. This suggests that mental model validation occurs in System 2 (see for example Norman et al., 2017). System 1 (or the first mind) is subject to the economy principle, which means that implicit knowledge and belief-based pragmatic reasoning are quickly accessed but there is an innate desire to avoid more effortful reasoning and engage the formal logic common to System 2. As Keith E. Stanovich and Richard F. West (2000) argued, individuals have a propensity to violate almost every rule that would lead to a valid mental model and a plausible conclusion when engaged in their routine patterns of thinking:

> people assess probabilities incorrectly, they display confirmation bias, they test hypotheses inefficiently, they violate the axioms of utility theory, they do not properly calibrate degrees of belief, they over-project their own opinions on others, they allow prior knowledge to become implicated in deductive reasoning.
>
> (p. 645)

It is reasonable to conclude, then, that System 2 reasoning represents a higher probability of validating an accurate mental model of reality than does System 1 reasoning. However, the probability of validation is perhaps incalculable. Deductive reasoning might be the basis for calculating such a probability, but it is beset with an array of nebulous variables such as the degree of uncertainty and complexity in the problem, the applicable knowledge held by the user, and how the user uses that knowledge. Moreover, there is no consensus as to what triggers the jump from System 1 to System 2. One theory, the 'MODE model' (Motivation and Opportunity as DEterminants of the attitude–behavior relation) (Fazio, 1990; Fazio & Towels-Schwen, 1999; Olson & Fazio, 2012) which centred on the relationship between attitude and behaviour, argued that the foremost motivating force to engage more effortful reflection is the human aspiration to be accurate and reach a valid conclusion. In short, the fundamental human need to be proven right.

This followed from social psychologist Arie W. Kruglanski's (1989) Lay Epistemic Theory (LET), which postulated a general need for cognitive closure that in turn leads to the acquisition of new knowledge and thus knowledge formation. Kruglanski identified two types of cognitive closure: non-specific and specific. 'Non-specific closure' refers to the urge to seek the answer to a question when the user is faced with ambiguity and perplexity. In this instance, any answer will suffice and the cognitive process is fundamentally intuitive. 'Specific closure' refers to the quest for a specific answer to a question and as such requires deeper more reflective thinking and reasoning. A prevailing factor as to which one was prioritise was the time constraint in which to validate a hypothesis. Fear of reaching an invalid conclusion also reduced the urgency and thus allowed for more effortful reflection.

With respect to Russel H. Fazio's (1990) MODE model, I submit that the desire or need to validate an accurate conclusion is more associated with the provisos of one's vocation; scientists and other professionals such as engineers and surgeons for example, rather than an innate desire to achieve as much when confronted with the everyday and a disposition for cognitive economy. Kruglanski's (1989) reference to the 'fear of invalidity' of a conclusion aligns more with my story herein. More precisely, that the jump to more effortful reasoning – from System 1 to System 2 – is prompted by a general negativity bias (discussed in detail later).

From the foregoing discussion it can be construed that past experience would allow an individual to bypass more effortful thinking and jump straight to a plausible conclusion. 'We remember the past while imagining the future' (Kaku, 2014). Henry Markovits and Pierre Barrouillet (2002) concurred as much by noting that transient mental models, constructed by retrieving relevant knowledge from long-term memory, are increasingly stored in working memory as capacity increases. For example, a brain surgeon, presumably, does not need to periodically pause mid-operation to consult a 'how to do manual' or reflect on the next procedure. The procedure is quickly drawn from a set of explicit mental models – a logical and systematic process finely tuned to the demands of state-of-the-art brain surgery. One would hope that to be the case anyway.

It is apt here to consider the argument that human behaviour is driven by a fusion of genes and experiences (nature and nurture), and thus renders the 'free will' to deal with certain issues, a redundant enterprise. Environmental psychologist Robert Gifford (2011), has argued that the human brain has hardly evolved over thousands of years. It remains redolent of primal times, including a dependency on agriculture, immediate dangers, exploitable resources and the 'here and now'. As a consequence, it is naturally inconsistent with the problems of modern society, such as climate change, which is slow and usually perceived to be distant. Gifford further argued that our fundamentally ancient brains are capable of dealing with climate change but this would not happen without considerable effort. Although the new brain beckons, the old brain remains firmly entrenched. There is an obvious next question here: how do politicians and policymakers tackle contemporary slow-moving phenomena like climate change if they are using a primordial brain?

Thus far we have discussed the theory of two minds in one brain. Now consider, in accord with psychiatrist and neuroscientist, Peter C. Whybrow (2015), that the human brain is an evolutionary hierarchy of three brains in one. Our bodily functions and quest for survival are controlled by our reptilian brain. This brain focusses on the primal reward pathways of desire, pleasure, curiosity, and short-term survival. Throughout the course of millennia, the early mammalian brain – the limbic cortex – developed around this primitive core. The early mammalian brain embodies the social behaviours of kinship and the nurture of the young. The continuous expansion of this cortex culminated in the emergence of *Homo sapiens* (our species: *Homo* [man] and *sapiens* [wise]). Although historical records are constantly being updated, *Homo sapiens* are believed to have originated in sub-Saharan Africa around 200,000 BC. The exceptional growth of the orbital-frontal region is what makes humans distinct from other primates. It is significantly more advanced in humans than it is in other animals. The prefrontal cortex of the human brain is where sophisticated thinking and reasoning occurs. The crux of Whybrow's hypothesis is that modern humans rely on a brain that is only marginally adapted to the challenge of long-term survival. Thinking beyond our own generation is a most difficult proposition.

In contrast to the dual-process theory, but more attuned to Whybrow's hypothesis, Robert E. Ornstein and Paul R. Ehrlich (2011) attributed many of our finest achievements in the contemporary world to 'a refinement of the old mind'. They did, however, concede to its limitations:

> The human mental 'hardware' – our senses and brains – is effectively fixed. That hardware equips us with what we call the old mind. Although we are evolving, our mental machinery will not change biologically in time to help us solve our problems. The same mental routines that originally signalled abrupt changes in the old world are pressed into service to perceive and decide about unprecedented dangers in the new.
>
> (p. 11)

As a matter of urgency, is there a way to speed up this evolutionary process to avoid the possibility of another mass extinction? The idea of 'maladaptation' portends to the theory that modern humans are fundamentally adapted to the Stone Age (the Pleistocene epoch, or thereabouts), rather than the world of today. Evolutionary time lags suggest that because evolutionary change requires millennia, humans are inadvertently designed for the previous environment – we are thus more than just anachronistic but positively ancient. In short, 'we carry around a Stone Age brain in a modern environment' (Buss, 2008, p. 19). As a consequence, Mars® Bars make us fat and the scourge of diabetes ravages the human population.

Diabetes, a by-product of our capitalist constitution, is the downside of a sugar-coated Western way of life. Studies reveal that, at the neurobiological level, the addictive powers of the sugar reward are ostensibly more robust than that of cocaine (see for example Ahmed, Guillem, & Vandaele, 2013). Sugar is the lifeblood of the 'get fat fast' corporate sales pitch. The World Health Organization (2014) has estimated that worldwide more than 347 million people have diabetes which is closely linked to obesity. About 90 per cent of these suffer from Type 2 diabetes: Type 1 diabetes is the body's lack of insulin production; Type 2 diabetes is the body's ineffective use of insulin. A healthy insulin system regulates the body's glucose (sugar) levels. Abnormally high blood sugar levels damage organ systems and, if untreated, lead to premature death. It is forecast that fatalities from diabetes will increase by 50 per cent over the next decade.

Of course, the main problem with this is the obvious one. Although Mars® Bars do make us fat if we consume too many, humans, by all accounts, are the most successful large organisms in the history of life on Earth (Dupré, 2012). *Homo urbanis* has bloomed spectacularly. Though climate change might dampen his and her prosperity, there is no inkling of an end to exponential population growth. Moreover, in defence of a 'sweet tooth' and climate change inaction, there is no absolute evolutionary premise that would guarantee the adoption of a particular optimal adaptive trait that seemed like a good idea at a particular time along the evolutionary highway of humankind. Adaptations might turn out to be good, or not so good, in any given case. Humans are thus also the luckiest large organisms to have populated the planet. Although, 'go forth and populate', might be the poison pill that ultimately brings the human species undone. The 'selfish gene' proving too selfish and natural selection too successful.

Although it is generally accepted that natural selection cannot predict the forthcoming environment, adaptations that are useful, as such, have a genetic basis (Buss, 2008). Is it plausible, then, that mental-modelling is simply an adaptation that makes it possible to anticipate the forthcoming environment through the transmission of a particular gene? Better still, we might reduce evolutionary adaptation from millennia to minutes by artificially adopting a System 2 mind mode expeditiously. Could it be that simple? Or is it that we are necessarily chained to the present because that is where the immediate danger lies? Survival in the first instance will, presumably, always override long-term survival. This is the genetic algorithm that delivers the highest probability of survival. However, a slowly unfolding catastrophic event renders this algorithm redundant.

The constructionist approach views the person and the environment as a single unit. Perception (sense perception) is made possible by a background of personal knowledge and place. This gives it acuity and direction. Users derive knowledge and the meaning of things from their experiences with them. For example, studies by Adam J. Fay and Jon K. Maner (2014), found that low-level sensory experiences interact with features of the person and the situation to shape social cognition and behaviour. Moreover, Mesquita, Barrett, and Smith (2010) argued that mental states and human behaviour are a consequence of moment-by-moment interaction with the environment, rather than proceeding in some independent, invariant, and content-free manner from preformed dispositions or causes. A mind exists in context, referred to as the 'context principle'. Insofar as the 'brain is plastic, dependent and malleable based on different environments, experiences and social situations, confirms the importance of the environment and social interactions in brain function and learning' (Hall et al., 2014, p. 74). As psychologist Thalma Lobel (2014) pointed out, it is disconcerting to discover that seemingly tangential environmental factors and physical sensations continuously influence our behaviour. The human brain is ecologically oriented but the body is seemingly unaware of its state of mind.

How quickly or easily an image is pieced together might depend on the physical disposition and place at the time (a disposition to behave in a certain way in keeping with the immediate environment and present company) and familiarity based on prior experience (experiential or general knowledge and working memory capacity). These are subject to interpretation and conceptualisation and to discourse aptitude. Language (the faculty of discourse) is rudimentary to the comprehension of the reality of the everyday. It is the foremost 'single symbolic form' used to express oneself and convey a communication (Willis, 2013). However, as argued by Johnson-Laird (2013), mental models are not constructed in response to the linguistic structure of discourse but in response to the perceived reality that discourse brings to mind – syntax as opposed to semantics, as previously discussed – hence, the ubiquitous human capacity to misunderstand.

Western philosophy imparts that we must understand something in order to give it meaning; or by giving it meaning we somehow profess to understand it. We are compulsive 'meaning-makers' (Valsiner, 2014). We have a classifying and cataloguing obsession. Our culture (Western culture) is long accustomed to splitting and dividing all things as a means of control (McLuhan, 2013). Only by atomistically narrowing it down can we say that we truly understand and will thus be able to exert control over whatever it is we have come to understand. From this well-ordered Western perspective, we readily perceive things that are well-documented and well-explained. In contrast, we struggle with abstractness. Abstractness is a cognitive complexity that befuddles endeavours to comprehend the sum of its parts. For instance, how does one translate and convey the abstractness of climate change so that it makes enough sense to initiate a costly and deliberative action? Is language, which has been the dominant medium, the best and in reality, only mode?

Lipari (2014) observed that language had been carved up into three separate processes of thinking, speaking, and listening, 'but was rather an organic whole, a confused multiplicity that echoes, vibrates, and pulses in nonlinear time' (p. 137). In essence, humans have been pre-conditioned to understand something because it has order, not because it has implications. Implications are future-oriented and fuzzy. Climate change has implications but no discernible order.

It is everywhere but nowhere in particular. Consider the folly, however, of failing to implement deliberative climate change policy simply because humanity failed to understand climate change because it did not pass the normative test of understanding. It has no order and no obligation to adhere to any such criteria.

Lipari (2014) argued that understanding and misunderstanding are inexorably linked. Misunderstanding moves us closer to understanding, and understanding is always partial and incomplete – they are at once both 'poison and remedy'. According to Z. D. Gurevitch (1989), one can only truly engage in a dialogue if able to cast off previous understandings, or preconceived images. This does not entail accepting something as true belief, but a conscious endeavour to open one's mind to the possibilities of 'not understanding rather than understanding'. Lipari (2014) wrote that 'there is great strength in not understanding – in giving up our convictions and certainties to let understanding evolve' (p. 139). No less than a call to liberate oneself from habitual thinking. To reimagine the mental models we hold into more plausible conclusions. The idea of letting go of false premises whose only purpose was to provide some 'peace of mind' and satisfy our inherent need to understand. In light of this, a commensurate policy response to climate change might commence with our capacity to be comfortable with the notion that we misunderstand it. To cast out preconceived mental models of the climate change object and start anew.

This might be easier said than done. Perhaps misunderstanding or understanding is simply undesirable? The half-truths of political speak might cease to resonate with the publics, inspiring them to literally tune out, switch off because 'we've heard it all before'. They might simply go through the motions of nodding and listening but in reality, not listening at all. As Lipari (2014) observed:

> The cognitive structures of language are particularly dangerous when we become so habituated to them that they are automatic and invisible to us. They distort our perceptions and point us to see what we expect to see and ignore everything else. We become habituated to the familiar – we listen to others on 'automatic pilot,' hearing only what we expect to hear because we don't listen to what is actually before us.
>
> (pp. 98–99)

But in the case of political rhetoric, is there anything new before us? Does tuning into the same old political rhetoric make any sense at all? Indeed, a clear and concise answer to a question is more of a revelation than an expectation. Answers are often not answers at all but cynically reframed retorts aimed at lambasting the opposition party. The publics are, in essence, reduced to reading between the lines in order to achieve some semblance of what is being communicated. Political speak rarely equates to the action taken. Generally speaking, the organising principles of the political mind are primal and bent on surviving the 24/7 media cycle. Political speak is invariably unsophisticated, unimaginative, and misaligned with that of the publics' expectations. Expectations that have periodically held great hope only to be thwarted by political indecision, indiscretion, obfuscation, and the duplicitous framing of important issues like climate change.

The reality of an object, scenario, or phenomenon, as a consequence of discourse (thinking, speaking, and listening) and imagination, might not match the expectations created by one's mental model once one arrives at that point in the future – 'this is not what I pictured'. The same applies to our political leaders. Expected utility falls short of realised utility. This could entail the promise of a visionary political leader or a meal at a restaurant that falls short of expectations. The option then is to adjust the way in which one interprets the image and reimagine another scenario that is more amenable in terms of expectations. Alternatively, one might just choose to lower his or her expectations. The expectation, for example, that pre-election promises will probably not match post-election policy. Expectations 'guide how people encode (attend and interpret), remember, and respond (judge and interact) in their social worlds' (Fiske, 2014, p. 156). Expectations, given an inkling of optimism, inspire one to imagine that anything is possible.

As Einstein declared: 'Imagination is more important than knowledge. For knowledge is limited, whereas imagination embraces the entire world, stimulating progress, giving birth to evolution' (as cited in Calaprice, 2000, p. 7).[2] Einstein thus inferred that one could be knowledgeable but lack imagination. And without imagination one is limited in one's capacity to fully exercise one's knowledge. John Cocking and Penelope Murray (2013) argued that 'Imagination, once regarded at best as a useful intermediary between sensation and reason, at worst as a dangerous and delusive power, has now become the prime faculty of the human mind' (p. vii). They described imagination as not simply a reproductive mechanism but a wayward and irrepressible faculty capable of changing our perceptions of the world – our worldview. Furthermore, in addition to reproducing mental images of sensible appearances, the imagination is adept at playing around with sense impressions by manipulating them to produce images of imaginary things. Renaissance philosopher, scientist and statesman, Francis Bacon (1605), taken from an interpretation by Cocking and Murray (2013),[3] described imagination this way:

> imagination carries messages from the senses to the mind, messages which reason interprets to provide understanding; and that it also acts as an intermediary between the decision to act in a certain way and the practical carrying-out of the action, making as it were an image of what is to be done, which image or mental picture is then translated into concrete action, inserted in the real, sensible world of space and time.
>
> (p. 268)

Bacon (1605) construed the imagination as a conduit to the mind which transmits messages that are interpreted by reason. He also provided an apt description of a mental model: 'making as it were an image of what is to be done, which image or mental picture is then translated into concrete action'. The imagination, then, is an internal mechanism; a graphical interface (see Figure 6.1). It interprets and conceptualises sensory data – by way of a reasoning process – that is collected by the sensory organs and through recollection and introspection. The imagination is the mechanism that interweaves external and introspective data to construct an image of all the possibilities. The image is constructed in accord with the organising principles contained in each individual's cultural codebook. The mental model formed provides a blueprint for the decision to be made and the action to be taken. The end result is the social construction of one's reality; a juxtaposition of faithfully reporting the reality of the everyday while at the same time bending to the imagination that seeks to transcend the everyday.

That part of the cultural codebook, the historical path (see Figure 6.1), is important because it defines the chronological and predictive qualities of mental-modelling. It is chronological in the sense that more recent experiences exert a stronger influence over the imagination and thus the image formed. It shapes the present through memories of the past and establishes a common frame of reference when projecting future actions. For instance, one might say that right-wing governments have established a history of climate change conservatism which is predictive of future decision-making and action. The mental model formed then determines what decision to make and what action to take based on historical and future reference points.

Why is culture so important and what effect does it have on the mental model formed? Hall et al. (2014) argued that 'The infrastructure of the brain is shaped by and shapes culture' (p. 47). People live in culturally specific ways – 'culture provides form and shape to the self ['self' is an entity that supposedly stands separate from our minds and bodies] to the extent that the relevant social world is arranged and practiced differently' (Zevnik, 2014, p. 17). Moreover, cultural neuroscientists theorise that culture's influence on the brain results from repeated participation and engagement in culture's conventions, routines, and socially shared scripts for action to the extent that the brain reflects culture. Hall ([1959] 1980) understood this decades ago when he wrote of his countrymen's exploits abroad: 'Honest and sincere men in the field continue to fail to grasp the true significance of the fact that culture controls behaviour in deep and persisting ways, many of which are outside of awareness and therefore beyond conscious control of the individual' (p. 25).

It stands to reason that if different cultures produce different thinking processes that are built into the human brain, a culture of denialism would have the same effect. Apart from the well-documented differences in values and beliefs, a cultural mindset that is deficient or contaminated – it has acquired certain mindware gaps that make it predisposed to the denial of phenomena such as climate change – even though the denied phenomena are empirically stated, might be more than just a case of illogical or irrational reasoning. It might manifest as a physical anomaly. The 'brain is the quintessential biological organ' (Kitayama & Park, 2010, p. 125) and prolonged and repeated experiences can cause functional and structural changes to it (Maguire et al., 2000). 'Functional and structural changes' presumably refer to how the brain is organised; however, the literature is not clear as to the extent of this.

The truth hurts, literally. According to Heejung S. Kim and Joni Y. Sasaki (2013), a violation of an individual's cultural meaning has health implications in the form of cardiovascular and neuroendocrine irregularities. The truth might therefore hurt, both in a mental and physical sense. Updating to a new mental model that strongly contradicts the culturally inculcated one, might thus cause ill-health. Not to mention that a structural change to the mind/brain, if visualised as a physical change, prompts one to imagine a rather painful experience. The idea of structural differences of the mind can be traced back to German philosopher Immanuel Kant. In his *Critique of Pure Reason* (1781), he argued that the human mind is not a blank slate that inscribed observations from the outside world (as cited in Pirie, 2009); rather, the human mind is an active agent that participates in acquiring knowledge by processing the inputs it perceives which means that our knowledge of the world rests on the structure of our minds. I surmise that this structure is premised on the organising principles that comprise one's cultural codebook.

With respect to this, a denialist culture could theoretically extend to the culture of a country and the structure of its collective mindset. For instance, the Australian culture and the colloquialism of 'she'll be right mate' – a positivity bias as opposed to a negativity bias (discussed in more detail later) – might shape the country's perception of climate change. Australians are a conflation of Jeremy Bentham's utilitarian ideals and the 'pie-eyed optimism' characteristic of the baby boomer generation (see for example Offord, Kerruish, Garbutt, Pavlovic, & Wessell, 2014; Riggs & Turner, 2000) – whatever the problem 'she'll be right mate'! The idea that 'everything is OK' is reinforced as a cultural norm, which might usurp scientific reality. As Ronald Conway observed in his book *The Great Australian Stupor: An Interpretation of the Australian Way of Life* (1971):

> the Australian mind ... is certainly deficient in the tragic view of life.... The point that suffering and crisis may throw some clearer light on the design of existence, that it might be the way the mindless ones are supposed to reflect and grow a little before they die – this is a proposition to which so many Australians are likely to retort: 'bloody lot of rubbish', and stalk irately away.
>
> (pp. 256–257)

A self-proclaimed 'Aussie' culture of sanguinity might also be interpreted as value placed on unsubstantiated belief and unrealistic optimism. One might then surmise that the great Australian ethos of 'she'll be right mate' is possessed of crystallised inhibitors and as such, is the antithesis of System 2 reasoning. More to the point, in reference to both human action and human health, 'People need to be optimistic about the future' (Jones, 2015, p. 250). But what might be good for one's health might be devastating for the health of the planet. This has implications for reflective thinking and reasoning, especially with respect to negative stimuli like climate change. For instance, there is a fine line between being overly optimistic and unduly pessimistic. The expression 'even paranoids have real enemies' perhaps best defines this demarcation (attributed to Delmore Schwartz or Henry Kissinger or Sigmund Freud, among others, but probably anonymous). A culture in which optimism overrules a perilous encounter with Mother Nature, to the point of nonchalance, is an invitation to disaster.

Research has revealed that culture is not necessarily a stable phenomenon – there is a capacity for variation and change in an individual's and a community's practices. Barbara Rogoff (2003) asserted that 'people develop as participants in cultural communities. Their development can be understood only in light of the cultural practices and circumstances of their communities – which also change' (pp. 3–4). One might surmise, however, that acquiring a new set of organising principles is not such a straightforward proposition. There is an inherent desire to retain the familiar – long-held mental models that are not so malleable.

At the most basic level, we are, in essence, products of our time and place: 'When we act, we act as a historically located individual, creating and created by personal and shared histories of participation and reification' (Hall et al., 2014, p. 43). According to educational theorist Etienne C. Wenger (1998), 'reification' is a product or process whereby abstractions, or ideas, are given an independent existence, or concreteness, of their own. Abstractions then become meaningful and shape one's experiences. Reification is how 'we impress our meanings onto the world, and then, on meeting them again, perceive them as existing as external facts' (Hall et al., 2014, p. 205).

Reification has implications for climate change. The nature of climate change is abstractness, which allows it to more easily assume an identity independent of its scientific reality. Similarly, the idea of climate change shapes the experience of it because humans might have yet to physically experience it – or recognise that they experienced it – much like Plato's allegory of shadows on a cave wall. 'the many live peacefully in their cave among themselves, mere spectators of images, involved in no action and hence threatened by nobody' (Arendt, 2005, p. 296). If the occupants are forced or motivated to emerge from the darkness of Plato's cave there is the harsh light of reality to contend with – a reality that might require a deliberative action.

Reality is relative. As pointed out by Hawking and Mlodinow (2010), 'These mental concepts [mental models] are the only reality we can know … a well-constructed [mental] model creates a reality of its own' (p. 172). People create their own meaning from their encounters with the world and then behave in ways that are consistent with those meanings (although not the same, meanings are analogous to beliefs). But as noted by Wenger (1998), although reification is ongoing and enriching it can also be misleading. Giving our own projected reality to an abstract object, scenario, or phenomenon, means that the object, scenario, or phenomenon itself, might be totally detached from reality. That is, 'reification suggests that forms can take on a life of their own, beyond the context of their origin' (Wenger, 1998, pp. 61–62). These forms or artefacts are thus internalised and infiltrate our systems of thought.

Foucault ([1989] 2012) argued that the creation of systems of thought were not intentional but the product of 'thinking and acting' which presuppose a particular pattern of knowledge. The pattern of knowledge then becomes typical of a historical period (the Anthropocene, for example). Moreover, according to Foucault, historical events are outcomes of disorderly and complex social interactions – ways of thinking and acting – and not orderly and sequential as a linear chronology would suggest. Conversely, the Italian philosopher Giambattista Vico, who in his publication *First New Science* (1725), held that history was cyclical. Civilisations and cultures progressed through sequential phases of development and a set historical pattern emerged: 'the history of all nations will develop in accordance with an identical pattern of social, political and cultural change' (as cited in Pompa, [1725] 2002, p. xxv). The reason being is that they all share the same nature and progress on the basis of an increasing rational understanding of the true nature of things. In essence, the 'circle' that is culturistic of Eastern thought (Foucault's hypothesis) as opposed to the 'line' that is culturistic of Western thought (Vico's hypothesis). And by extension, the circular mental model as opposed to the linear mental model.

Foucault ([1966] 1970) held that humankind's evolution was largely unpredictable and discordant in terms of the evolution of time and more contingent on the particular activities of people that pre-empted discovery and innovation. Thus, in concurrence with this reasoning, the golden ages of Classical Greece and the Age of Enlightenment were punctuated by the Dark Ages for no other reason than humans mental-modelled it, for whatever reason, and then collaboratively acted it out. Contrary to both Foucault and Vico, political philosopher Eric Voegelin ([1956] 2001) contended that history becomes intelligible when viewed as a struggle for order. And in the midst of that struggle, history can move backwards, forwards, and sideways. This ostensibly contradicts the contemporary metaphor of unending 'year on year' economic growth and an almost preordained pathway to the technological singularity. Ironically, when Vico first published his hypothesis in 1725, the argument later propounded by Foucault was in fashion. Vico's hypothesis was thus originally admonished. The question is: which hypothesis will hold true in the case of climate change? All three illustrate the fact that humanity remains in two minds as to whether the current era of the Anthropocene will inevitably continue or the present-day thinking and acting of humans will bring about a spectacular reversal, or perhaps, the middle ground of a sideways shift.

In his endeavours to explain why humanity devolves itself of one episteme – a consensus as to the principles that sanction the creation of valid knowledge – for another, Foucault ([1966] 1970) cited an emergent discontent with prevailing social structures and traditional perceptions of the world: 'words, classes, and wealth will acquire a mode of being no longer compatible with that of representation' (p. 221). A challenge to existing worldviews and ideas would emerge which in turn prompt a reimaging of our understanding of the world. It is unclear, however, as to what triggered these sometimes abrupt, or equally slow, shifts in worldviews. And as Galileo would probably attest to if he were alive today, they often come at great personal sacrifice: the abandonment of what was most familiar for something new and unfamiliar. Perhaps Heraclitus explained it best by declaring that 'nothing endures but change'[4] (as cited in Huskinson & Stein, 2015, p. 1). Accepting the fact that it just happens as a product of time itself and for no other logical reason.

Delving further into Foucault's (1970 [1966]) hypothesis, one could construe that historical events are the outcome of collaborative mental-modelling: the deconstruction and reconstruction, or decomposition and composition, of cognitive representations that unconsciously and consciously precipitate change. Discontinuity is a function of the end-state of cognitive collaboration: of continual decay and renewal. Decay is perhaps easier because it comes naturally. Renewal is more painful because it means surrendering to decay with no guarantee of renewal. Considering this, is climate change a subliminal prompt for the renewal of the human spirit? Not just in a political, economic, scientific, and social sense, but also in a metaphysical sense. For instance, what is the nature of climate change? Is it a consequence of human inventiveness or human destructiveness? If inventiveness ultimately leads to destructiveness, a circular chronology in which history is destined to repeat itself, then man is the inventor of his own destruction. In consideration of this, there is the possibility that because of its many parts and effects, the object, scenario, and phenomenon of climate change, might not still be able to be comprehensively imagined.

To believe or not to believe: that is the question

Stephen W. Hawking (1988) in his book *A Brief History of Time*, concluded that there can be no God because time did not exist before the 'big bang'; hence, there was no 'time' in which God could create the universe. Furthermore, the big bang did not require God's participation and could be explained by 'spontaneous creation' (Hawking & Mlodinow, 2010). Conversely, believers in God could just as easily argue that God is not subject to time. Faith is all that is needed. Under the condition of faith, belief is both incontestable and inescapable. It quenches an insatiable thirst by giving meaning to the world in order to make one's life bearable if not the Good Life. Reality might be relative but belief is relevant.

For instance, I once offered a good friend of many years a copy of *The God Delusion* by Richard Dawkins (2006a). Dawkins constructed a rational argument of why God could not exist. He painted the belief in God as a mass delusion: an irrational false belief heavily primed by social pressures and programming. My friend, a dedicated Christian, rejected the book on the grounds that it was irrelevant. The gist of his response went something like this:

> Why should I listen to an argument, however rational, when my relationship with God has delivered me through the hard times of two broken marriages and the periodical estrangement of my children? Being Christian and believing in God helps me negotiate the travesties of life … solemnity and solace is found within the Christian group to which I belong.

It is difficult, no near impossible, to argue with this reasoning. As philosopher Paul Weiss (1946) pointed out: 'To reach a state of complete confidence one need do little more than protect a belief from the threat of criticism by eliciting the support of tradition or of gross observation' (p. 132). Religion is certainly a benefactor of gross observation. People have always believed in some form of divine being … billions of people around the world believe in God – hence, *argumentum ad populum*. A proposition is true because the majority, or a large number of people, believe it to be true. Likewise, the weather has always changed … this summer is more or less hotter than last summer and the summer before that. This we have observed to be true. Similarly, if climate change is accepted as reality and perhaps an end to life as we know it, is it possible to accept God as the imperfect creator? Then again, if climate change is a divine intervention brought about to curb overpopulation that threatens human existence, then God is our saviour. The Lord works in mysterious ways. Besides, if there is no God then whom or what are we praying to?

I surmise however that religion, stripped of the cloth, is just another ideology – albeit a very powerful and influential one. In my friend's case, his 'belief system' constitutes an important resource. But it is both enabling and disabling. Enabling because it can be called upon for spiritual sustenance; disabling because it has the potential to close one's mind to other possibilities. For instance, as Gifford (2011) explained: 'Some people take little or no climate-related action because they believe a religious deity or Mother Nature (as a secular deity) either will not forsake them or will do what it wishes anyway' (p. 293). Indeed, religion is a profoundly powerful thing. People will sacrifice their lives to preserve their religious rites, their language and literature, their cities and their art. They might do this to the extreme of denying the same rights for others who pursue an alternative path.

Humans are apt to choose the mental model that best helps them through the day. From this worldview, whether God truly exists is irrelevant. Similarly, to the vast majority of people in a crowded city street confronted with a growing ensemble of everyday obstacles, climate change is simply irrelevant. As argued by Peter L. Berger and Thomas Luckmann (1967): 'everyday life is structured in terms of relevances', determined by a conflation of immediate pragmatism and social circumstance (p. 59). Belief is relevant. Religious beliefs might be more relevant and thus provide greater benefit than other kinds, such as those imposed, unexamined, or tenuous. For instance, there are those bodies of knowledge apart from the vast array of religious knowledge and the body of empirical knowledge that come to be socially accepted as reality. Like 'year on year' economic growth and exponential population growth are sustainable. Moreover, if knowledge is imposed by others and unexamined by the user, then it is, in effect, taken as reality. The person in the street is not ordinarily concerned with what is real and what is not: unless of course he or she encounters an obstacle that demands that he or she does. In truth, unless one seeks to find out or is indeed forced to find out, there is no urgency to establish where validated ends and invalidated begins. Climate change might assume this quality if it has not been experienced physically, or recognised that it has been experienced physically.

According to Frederick F. Schmitt's (2014) interpretation of prominent English philosopher David Hume ([1739–1740] 1978), a belief is more vivacious[5] than a mere conception (for example, the belief that the Earth will always right herself has more vivacity than the conception that climate change poses some abstract curse on the wellbeing of humankind) and constitutes a 'practical natural function' that serves to motivate behaviour to avert calamity. If a belief averts calamity – this is not a question of whether God exists – although it might not be true (as in the scenario foregoing in brackets) it has obvious appeal. As Markovits (2014) explained: people commonly reason in what appear to be totally illogical ways (see also Wason, 1968), and rationality and intelligence are two different things. This means that one can be both intelligent and irrational at the same time. Think of Ted Kaczynski, the Unabomber, an exceptionally intelligent individual prone to exceptionally irrational behaviour.[6] Stanovich (2009) termed this seeming paradox 'dysrationalia': 'the inability to think and behave rationally despite having adequate intelligence' (p. 7). From this perspective, who is to say that the rationale of the climate scientist or the political practitioner, at any given time, is superior to that of the armchair thinker?

Nevertheless, there is an expectation that intelligent people will act rationally, and when it is perceived that they do not, trust in them can instantly evaporate. Consider for instance the 2009 'Climategate' affair that led to a waning of public trust in climate science and scientists per se. Climategate was the online release of more than 1000 emails and other documents without their authors' consent. The contents of which showed some climate scientists in a biased, self-serving light which threatened to undermine the credibility of climate science and climate change reporting. The emails implied that British and American scientists had conspired to quash research they disagreed with, and deliberately alter their research results so that global warming would appear much worse than it really was.

In March 2010, Professor Philip Jones, director of the Climate Research Unit (CRU) at the University of East Anglia (from where the emails originated) and deemed to be a central figure in the Climategate affair, was exonerated of any wrongdoing by a British parliamentary committee. The idea of climate change, however, remained under sustained attack by sceptics and denialists. Trust is thus transitory: it is most difficult to gain and can be lost in the blink of an eye. The publics are harsh judges when the issue of trust comes into question; which in turn, becomes a question of belief in someone or something, rationality notwithstanding.

Kahneman and Tversky (1973) described intuitive predictions as judgmental and subject to the 'representativeness heuristic'. Humans use the representativeness heuristic to predict the outcome that appears most representative of the evidence at hand. The consequence of this is that intuitive predictions might be oblivious to the validity of the evidence or the probability of the outcome that is in violation of statistical probability and thus raises the conjunction fallacy. Intuitive predictions are not based on a logical probability distribution, such as that achieved after many thousands of rolls by a finely machined dice in a true game of chance. The mind is simply not calibrated to that level of statistical perfection. Instead, according to Kahneman and Tversky (1973), 'people predict by representativeness, that is, they select or order outcomes by the degree to which the outcomes represent the essential features of the evidence' (pp. 237–238). The outcome, then, is subject to an 'unjustified confidence in predictions, and some fallacious intuitions concerning regression effects',[7] both of which can be attributed to the representativeness heuristic (Kahneman & Tversky, 1973, p. 238). Under these conditions, one could construe prediction as a function of an intuitive belief in the evidence rather than evidence that empirically justifies the belief. As in the case of Climategate discussed foregoing.

There is no consensus as to the epistemological status of both intuition and belief. The superiority of logical deduction over intuition is based on the premise that logic provides a 'more certain' worldview of the possible states of the world – but how more certain is unclear. Considering this, it is reasonable to suggest that in the same way that people do not normally assess the risk that confronts them based on the logic of probability, they likewise do not normally discriminate between intuition, belief, and knowledge, as they go about their everyday. For example, the IPCC (discussed in Chapter 1) strongly recommended that humankind should, without delay, pursue a course of greenhouse gas emissions reduction. Hence, there should be a normative compulsion to be well informed of the underlying science and take the necessary precautions and actions. A logical deduction would be: the Earth is warming. The warming is dangerous and is largely human-induced through the burning of fossil fuels; therefore, we must cease burning fossil fuels and look to clean, efficient and safe energy alternatives. On the face of it, an intelligent person arriving at any other conclusion might be considered irrational. Similarly, it would be immoral to accept the science and not implement a commensurate climate change policy response: to ignore the danger and free-ride on the efforts of others, or another nation, is positively unjust and unethical.

Just as irrational thinking and acting can coexist with a high degree of intelligence, so too can intuition, belief, and knowledge. One can be both a scientist and a Christian; a policeman and corrupt; a soldier and a loving wife; an environmentalist and a frequent flyer. Attitude and belief are not prescriptive of behaviour. As British polymath Bertrand Russell ([1926] 2014) intoned: the human mind was not designed to be imprisoned within the narrow walls of logical deduction and experiential fact: 'It is a dangerous error to confound truth with matter-of-fact. Our life is governed not only by facts, but by hopes; the kind of truthfulness which sees nothing but facts is a prison for the human spirit' (p. 85).

As such, belief and disbelief are equal opposites. Individuals might believe in a divine being in the same way they disbelieve in climate change. Religion, however, is perhaps the exception to the normative rule and thus not ideal in such an example. Although it is reflective, the cultural corpus of religious myths, rituals, and beliefs descends from prehistoric beginnings, constantly recurring in human thought and humankind's inexorable quest to comprehend and resolve its place in the world. The quest for 'the meaning of life' can be a lifelong distraction to living, notwithstanding an eternal longing for its resolution. Moreover, religion is philosophical but it is also innate and associated with the primitive mind. Indeed, 'delusions may be distortions of either intuitive or reflective interpretation' (Glover, 2014, p. 139). These links suggest that religious faith oscillates between System 1 and System 2 reasoning but has its roots in System 1.

For instance, it is presumed that Darwin agonised over publishing his theory of natural evolution for nearly 20 years because it contradicted his default mental model that was largely the providence of the Christian time and place. He understood the controversy that his theory would provoke in the eyes of the Church. Indeed, 24 years passed between his visit to the Galapagos Islands and the publication of his theory in 1859, *The Origin of the Species*, which was ultimately pressured by the imminent publication of a similar theory by Alfred Russel Wallace (cf. Wallace, [1855–1859] 2013). Although some authors have suggested that Darwin was delayed by frequent bouts of illness and his endeavour to iron out the finer details of his theory (see for example Wright, 1995). I suspect, however, that it was not so much Darwin's own Christian beliefs that caused his anguish – as a scientist his religious beliefs were irrelevant as his reasoning was attuned to System 2 – but that others might view his work as a betrayal of the Christian faith which was founded on age-old tradition and System 1 thinking.

It is universally accepted that the quest for knowledge is fundamental to humankind's evolution. But whether one is attuned to such knowledge has more to do with a general capacity and proclivity to engage a System 2 mind mode. More precisely, whether an individual's foundational orientation is manifested in System 1 or System 2. Acknowledged here is that religious belief and disbelief in the scientific consensus on climate change, although ostensibly both are forms of denialism, might be considered as two quite distinct sociological propositions. Independent scholar Herman J. Pietersen (2014), described a probable science-religion distinction as the ''Impersonal' (reflected in the search for abstract principles and laws of science and of metaphysics) and the 'Personal' (reflected in narratives of divinities and gods in the world)' (p. 1973). Although this distinction is important from a philosophical, political, and scientific perspective, it might not be obvious or meaningful to individuals as they go about their everyday. For instance, as with economic man and political man, is it possible for everyday man to shift and/or discern between the impersonal and personal in the course of everyday thinking and acting?

I suspect that religiosity plays an important role in thinking and acting with respect to climate change and further investigation is warranted. I also acknowledge that the science-religion debate is a highly specialised one.[8] As a point of interest I refer to a meta-analysis of 63 studies conducted by Zuckerman, Silberman, and Hall (2013) that indicated a significant negative correlation between intelligence and religiosity. They found that intelligent people were less susceptible to religious dogma. The analytic thinking (as opposed to intuitive thinking) that intelligent people tend to adopt served to undermine religious beliefs. Similarly, Sharon Bertsch and Brian J. Pesta (2009) reported that 'people who are less able to acquire the capacity for critical thought may rely more heavily on comfortable belief systems that provide uncontested (and incontestable) answers' (p. 232). Colette Mortreux and Jon Barnett (2009) highlighted the strength of belief systems and religiosity that can permeate a particular cultural context. Their research into the culture of the Tuvaluan people revealed the following:

> it is clear that religion plays a very significant role in shaping people's responses to climate change in Tuvalu. Of all the interviews conducted, around half raised religion in response to climate change. These people believed that climate change was not an issue of concern due to the special relationship Tuvalu shares with God and due to the promises God made to Noah in the bible. The strength of this belief is reflected in the national motto *Tuvalu mo te Atua*, meaning 'Tuvalu is for God, God is for Tuvalu'. In interviews people consistently referred to the story of Noah as evidence that God would not allow further flooding. There was a sense that Tuvalu was given by God to the Tuvaluan people and that God would ensure that this would remain the case into the future.
>
> (p. 109)

There is an innocence and optimism about such endearing faith – a desire, or even intrinsic need, to believe. It is difficult not to identify with the Tuvaluan peoples and join them in their endeavour. Faith, like perception, has claims to reality.

Much of the foregoing discussion has focussed on whether the mental models formed are true and genuine representations of reality. The accuracy or correctness of mental models is contingent on the kind of reasoning used, in addition to a myriad of other things such as imagination and knowledge. According to American philosopher Richard Rorty (1979), Western philosophy remains captive of epistemology; specifically, the conundrum of equating knowledge with its internal representations. 'The problem of the image relative to the assumption is that the image is, in itself, on par with the objects of perception' (Sartre, [1936] 2012, p. xvi). Assumption might pre-empt true belief, which is also an assumption in the sense that what we perceive is reality. The 'objects of perception' are just that, objects perceived in a certain personalised way thus, humankind's ceaseless philosophical discomfort with external reality versus internal representation. Rorty (1979) however, described this inimitable fascination as futile – a fanciful endeavour to catalogue the compartments of the mind:

> To know is to represent accurately what is outside the mind; so to understand the possibility and nature of knowledge is to understand the way in which the mind is able to construct such representations. Philosophy's eternal concern is to be a general theory of representations, a theory which will divide culture up into the areas which represent reality well, those that represent it less well, and those that do not represent it at all (despite their pretense of doing so).
>
> (p. 3)

French philosopher and playwright Jean-Paul Sartre ([1936] 2012) wrote of René Descartes's endeavour to solve the reality/representation conundrum by applying 'pure intellect' as a means of differentiating true images from false ones. It is unclear, however, what pure intellect actually is. In reality, there is no empirical method of knowing whether Descartes was correct, or anyone else for that matter. American philosopher and psychologist John Dewey (1927) described the reality/representation precipice this way:

> We forget that facts are only data; that is, are only fragmentary, uncompleted meanings, and unless they are rounded out into complete ideas – a work which can only be done by hypotheses, by a free imagination of intellectual possibilities – they are as helpless as are all maimed things and as repellent as are needlessly thwarted ones.
>
> (pp. 8–9)

Similarly, philosopher of science Paul Feyerabend (1981), argued that scientific theory is solely contingent on an interpretation of the state of affairs it described; whereas realism acknowledges that known and unknown influences will affect the observer in addition to the true state of affairs. In essence, 'what we consider to be 'reality' is only an approximation that the brain makes to fill in the gaps' (Kaku, 2014, p. 40), and 'our brains govern virtually every action we take' (DeSalle & Tattersall, 2012, p. 3). Our actions are thus presumably irrational unless the brain has reliable information to cogently bridge these gaps.

My intent here was not to detract from the validity of knowledge or give ground to unsubstantiated opinion, but to highlight the philosophical dilemma that confronts the scientific reality of climate change. Although not aimed at climate change, I assume that Dewey had the same in mind. He aspired to the notion that facts, presumably as they pertain to knowledge, tell only part of the story. In the same way, the fact of climate change tells only part of the story. The idea of climate change needs inspiration to achieve maturity, and maturity is most commonly learned rather than acquired instinctively. For instance, climate change simultaneously exists in two spaces: the medley of mental models in our minds and in the physical dominion that it occupies. Like many other fanciful ideas it needs a credible, or even imaginative path to connect the two. At present, for the most part, it remains as fragmented facts and data – aloof and unabridged – and thus an incomplete idea that lacks maturity.

From this discussion, one could surmise that the validity of deductions derived from mental-modelling is not the most important outcome, but rather how useful deductions are in the context of the everyday; for example, religious belief that helps one negotiate the traumas of life – something that cannot be dismissed lightly. This is not to discount the importance of truth as a remedy to the destructive forces of ignorance. But even if the deduction mirrors the empirical evidence it might be fundamentally useless, even detrimental, to the everyday of the user. In this sense, it might be denied on the grounds that it does more harm than good. As previously discovered, the truth can hurt, both mentally and physically, and it has no obligation to be merciful. And a 'person', as defined by the classic utilitarians John Stuart Mill and Jeremy Bentham, is the sum of his or her experiences – a pursuer of pleasure and shunner of pain. But what if dangerous climate change, a life-threatening truth, has never been experienced in one's lifetime and never really enters one's mind because of this.

The unconscious mind rules *in absentia*

In absentia, Latin meaning, 'In his, her, or their absence' (Speake & LaFlaur, 1999, p. 320). In times of critical importance, this might also apply to the conscious mind. The evidence is strongly in favour of the dominance of the unconscious mind over the conscious one, but it is not unanimous. Ben R. Newell and David R. Shanks (2014), for example, argued that much of the research demonstrated that behaviour is under the control of the conscious mind or can be otherwise explained without resorting to unconscious influences. In direct response to this, Bernacer, Balderas, Martinez-Valbuena, Pastor, and Murill (2014) suggested that Newell and Shanks had ignored the process of habitual decision-making that routinely prompts the transfer of decisions and actions between the domains of the unconscious and conscious mind.

Dutch philosopher Benedict de Spinoza ([1883] 1955) wrote: 'Experience teaches us no less clearly than reason, that men believe themselves to be free, simply because they are conscious of their actions, and unconscious of the causes whereby those actions are determined' (p. 134). Spinoza was referring to humankind's general ignorance of causation. This goes further to explain Newell and Shanks's argument foregoing. One might be conscious of one's behaviour but the reasoning underlying it draws a blank. Spinoza is, in effect, making reference to dual-process theory and System 1 (unconscious and intuitive) and System 2 (conscious and reflective) reasoning. In short, the unconscious mind rules *in absentia* of reason. Mindful that knowledge of the reason that underlies the cause, affects the action taken and thus the outcome.

According to A. E. Allahverdyan and Aram Galstyan (2014), confirmation bias is ubiquitous in the fields of psychology, economics, and science. Ben Goldacre in his book *Bad Science* (2009) demonstrated that our 'truth-checking apparatus' can fail. Clever people are fooled by five cardinal cognitive illusions (see also Taleb, 2004):

> (1) we see patterns where there is only random noise; (2) we see causal relationships where there are none; (3) we overvalue confirmatory information for any given hypothesis; (4) we seek out confirmatory information for any given hypothesis; and (5) our assessment of the quality of new evidence is biased by our previous beliefs.
>
> (pp. 47–50)

Another major constraint to the conscious mind is that thinking and reasoning, for the most part, is neither logical nor rational. This is because reality and rationality are relative, and belief is selectively relevant.

David McRaney (2012) pointed to automaticity as a prime suspect for our absence of consciousness: there is a stream of knowledge emanating from psychology and cognitive science that suggests that 'you have no clue' why you think and act the way you do, make the decisions you do, and your memory is mostly confabulation. Kaku (2014) posed a similar hypothesis: 'The brain is influenced by thousands of unconscious factors that predispose us to make certain choices ahead of time, even if we think we made them ourselves' (p. 342). In short, your unconscious mind rules in the absence of your conscious one.

One might also pose the alternative argument that System 1 (that is largely automatic) holds System 2 in check. In accord with a violent history and a competition for survival, humankind unconsciously ranks self-preservation as the overriding priority – a manifestation of the autonomic nervous system (ANS). System 1 overrides the probability of a delayed reaction as a precautionary measure that might otherwise be thwarted by a proclivity for reflective thinking when a quick reaction is necessary to protect or remove one from harm's way ('fight or flight'). But this is problematic when it becomes the norm in inappropriate situations. A habitual misallocation of cognitive resources occurs, which to some extent, is involuntary. This is similar in effect to the dominance of the left side of the brain (which tends to distort reality and confabulate) over the right side of the brain (which is more precise and artistic and is capable of abstract thinking). Or put another way, if the unconscious mind drifts too far in the direction of mindlessness, the conscious mind will drag it back to reality – hopefully.

Zombies and learning to think again

AI could make zombies of us all. Researchers suggest that by 2049 AI will write a bestselling book; and 120 years from now, all human jobs will be automated (Grace, Salvatier, Dafoe, Zhang, & Evans, 2017). God forbid if human ingenuity and imagination are ever reduced to the processing capacity of some form of artificial intelligence. The same would apply to ethical decisions and moral judgements. Humans thus become superfluous in the grand scheme of things (see for example Dietrich, 2017). Although self-subjugation, whether by choice or duplicity, has its merits, the thought of losing one's soul, and various other faculties, to the idiosyncrasies of a machine is positively unappealing. In light of such thoughts, an emergent topic of discussion is whether the Internet is affecting our brains. Is computer-mediated communication (CMC) via the Internet causing the zombification of living souls? Some researchers (for example Kuss, 2015; Kuss & Griffiths, 2015) consider the behavioural addiction to online applications as similar to substance-related addictions and have labelled it a 'twenty-first-century epidemic'.

Abnormal cyberpsychology is a relatively new field that studies mental health related to Internet Addiction Disorder; mobile phone dependency (momophobia, the fear of being unable to make mobile phone contact), and social media and online gaming dependency. Psychologist Ronald J. Comer (2013), explained that the study of abnormal psychology,[9] shared four common features: Deviance: being different from the norm and/or displaying extreme behaviour that might even border on the bizarre; distress: being unpleasant or upsetting the person(s) around them; dysfunction: unable to perform everyday tasks in a constructive way; and danger: a danger to oneself and others. I would imagine that most of us have at some time experienced at least the first three – hopefully not to any extreme. Moreover, all four are common human traits that are generally held in check. When they are not, and a person displays extremes of anger or depression at the threat of losing contact with his or her mobile phone, or being barred from online gaming, for example, then something attune to Internet Addiction Disorder is potentially in play.

We have now entered the first of phase of the zombification of society. Mesmerised by FOMO (fear of missing out), we cluster in the insular cul-de-sacs of social media with like-minded people and deliberate on like-minded topics of no real consequence. In his whimsical narrative Nicholas Carr (2008) posed the question '*Is Google Making Us Stupid?*': 'For me, as for others, the Net is becoming a universal medium, the conduit for most of the information that flows through my eyes and ears and into my mind' (p. 90). But at the same time Carr was concerned that the Internet might be altering his brain and not in a good way. For example, after hours of searching and surfing the Internet, he found it difficult to immerse himself in a book. Something he used to find effortless and enjoyable but now he became fidgety and anxious. It was increasingly difficult to drag himself away from the virtual world to read a book.

In the contemporary battle between technology and the brain, Pulitzer Prize-winning journalist Matt Richtel (2014) posed the following: 'Was technology [such as the Internet] no longer the slave, but the master? Was it overtaking our powers of attention?' (p. 5). Moreover, is the ultimate purpose of technology to solve the problems created by technology? Certainly, public and private institutions and the publics at large are anxiously installing technology to defend against the burgeoning incidence and audacity of cyber scams and attacks. But what is technology anyway? I suspect that it is deceptively emotive in the sense that all we know is that we want more of it – to an end-state wherein we need not think and act for ourselves. At the most basic level, however, perhaps our fascination with it derives from its likeness to magic as a thing that delights us, as inferred in Arthur C. Clarke's Third Law: sufficiently advanced technology and magic are fundamentally indistinguishable from one another.

In contrast to Carr's (2008) experience foregoing, digital historian David J. Staley (2014) argued that research suggesting that the Internet might be rewiring our brain's synaptic protocols (see for example Greenfield, 2015; Mills, 2014) was no cause for alarm. On the contrary, the Internet, as Staley explained, had the potential to enhance the analogical and associative thinking that enabled individuals to make connections between seemingly disparate objects and data points: 'The Internet has not dulled our minds but has instead unleashed this pre-existing, if undervalued, portion of our cognitive architecture' (Staley, 2014,

p. 15). I would add here that it is often 'analogical and associative thinking' that is the source of our more deficient mental models. That is, rational thinking cannot be reduced to mere associations between objects and ideas: logical connections, through rational inference, need to be made. There is another glitch. The virtual world is littered with untruths (Seife, 2014) and 'bad science' is pervasive in media (Goldacre, 2009). According to Ben Goldacre (2009), a medical doctor and science writer, bullshitters abound in media to the extent that science is commonly portrayed as detached from reality; groundless and incomprehensible and thus open to undeserved criticism. Truth-checking takes time and is, therefore, only for those with too much time on their hands. More than one or two clicks away is an unsolicited antagonism.

One could argue that institutions, opposing political parties and their inculcated mental models, have ceased to resonate with each other in a meaningful way. The publics have stopped listening as a consequence. Drowned out by a cacophony of bullshit, propaganda, slanderous exchanges, and political inertia, the publics have donned the headphones, left the building, and headed for the peace and tranquillity of the virtual world. 'Let me know when it's over' is now the catchcry of the masses. As pointed out by renowned neuroscientist Susan Greenfield (2015), the 'here and now', two-dimensional virtual world is 'so vivid and mesmerising' that it outshines the dull mundanity of the real world. And for a growing number of its residents, its immediacy and significance easily supersede its 'smelly, tasty, touchy' three-dimensional counterpart. Thus, for those who have chosen to make their home in the virtual world, does the framing of a political communication from the real world have any meaningful effect at all? 'Don't you get it … we're not listening anymore'! Quite simply, bored with the irrelevance and adolescence of mainstream politico-media speak it is too easy to reach for the headphones, flip the switch, and tune out. The result is a virtual mental model – a virtual reality constructed in a virtual world.

The downside is that even the virtual world has its own biases. Much like the brain's search for a relevant mental model, digital search engines such as Google, incorporate their own set of cryptic biases. The digital pages retrieved are based on Google's algorithms that are designed to boost online presence and thus revenues. So, what is presumed to be an 'objective tool for research, is actually a bridge that enforces its own set of modalities of connectedness, judgement and biases' (Kouppanou, 2014, p. 86). In effect, Google incorporates its own default digital mental model. Its aim is to transpose that mental model into the brains of its users, something it does very well. Ironically, personal customisation is assumed to counter this as digital technology can be easily modified to accommodate a user's preferences; however, this effectively means overlaying the search engine's biases with the user's biases in order to retrieve information framed in the author's biases. Furthermore, the virtual world is premised on immediacy. Reflection, which is central to a deeper mode of thinking and reasoning is naturally discouraged … the next online experience is but a click away. As duly noted by philosopher of education René Vincente Arcilla (2002), 'How can one have online experiences that stimulate reflection – if that technology is designed to manufacture experiences of immediacy that curtail reflection?' (p. 463). Reflection is the key to consciousness. Suffice to say, reflection is the antithesis of the zombie.

'Zombie' is defined in the *Oxford Dictionary of English* as '(Philosophy) a hypothetical being that responds to stimulus as a person would but that does not experience consciousness' (Stevenson, 2014, p. 9829). I define zombie as the mindless enterprise of human behaviour in its unconscious endeavour to simulate what it means to be human. 'Mindlessness' can be described as a 'cognitively dis-engaged, generally clueless, uncritical, essentially automatic responding' (Fiske & Taylor, 2013, p. 444). The word 'zombie' (or zombi) has its origins in Afro-Caribbean folklore but did not become mainstream until post the 1920s. In 1915 the United States invaded Haiti where the voodoo religion was practiced and the living-dead were cast in perpetual servitude.[10] Filmmaker George A. Romero made the zombie genre come alive with a series of movies beginning with the satirical horror classic *Night of the Living Dead* in 1968. Well beyond material consumerism, people had begun to consume one another.

A zombie, then, is identical to you and me but absent of any conscious experience. I presume that both you and I are conscious because I am writing this and you are reading it. And if we believe, to some reasonable extent, that philosopher-mathematician René Descartes's ([1641] 1948) dualism argument has substance[11] – that the mind and the body can exist separate from one another – then the zombie is indubitably conceivable (see for example Chalmers, 1996; 2010). That is, if, like many believe, that our conscious selves can exist in the absence of our bodies, then it is also conceivable that living, functional bodies can also exist void of any conscious experiences (Levin, 2017). So how does one distinguish the conscious beings from a room full of zombies?

With respect to this, I previously underscored a number of cognitive organising principles that alluded to an answer as to why we are all not stumbling about like a cohort of zombies but seemingly holding to a diversity of opinions, thoughts, and actions. Zombies have no need for organising principles and are void of consciousness; they are the antithesis of System 2 thinkers and reasoners. Could it be that our twenty-first century fascination with zombies – for example, *Shaun of the Dead* (Wright et al., 2004); *World War Z: An Oral History of the Zombie War* (Brooks, 2006); *Zombieland* (Reese & Wernick, 2009); *The Zombie Survival Guide* (Brooks, 2011); *The Art of Eating Through The Zombie Apocalypse: A Cookbook and Culinary Survival Guide* (Wilson & Bauthus, 2014); *Not Your Average Zombie: Rehumanizing the Undead From Voodoo to Zombie Walks* (Kee, 2017); and AMC Networks' long-running Emmy-award-winning television series *The Walking Dead* (Davidson, 2010) – is a subliminal reflection of the 'self' expressing its disquiet with the anonymity and drudgery of the everyday? Or otherwise the passive acceptance thereof and the opportunity to see parts of themselves reflected in zombie media. As Kim Paffenroth (2011) put it:

> With a monster that is so fully and banally human, one interpretation that will surely always be open and likely, is that zombies are us – not the mysterious

> reviled 'other', however that is constructed – but us, in all our hungry, grasping, mindless simplicity. We will always be our worst enemies, and the ones we can never fully eliminate.
>
> (p. 24)

'Ignorance is bliss'. This can be applied to both humans and zombies. Furthermore, in a politically correct world, zombies have rights too: the right to be unconscious for instance. Perhaps Socrates was unknowingly describing the zombie when he issued his famous decree: 'The unexamined life is not worth living for a human being' (as cited in Colaiaco, 2013, p. 177). Moreover, zombies are like-minded in their singular quest for flesh, whereas humans, particularly liberals and conservatives, are seldom like-minded about issues that concern the environment. Zombies might lack consciousness, and this might be said about politicians also, but neither lacks ambition. In the case of Australia's climate change inaction, for example, Garnaut (2011) inferred that the Australian publics are indeed mindlessly trailing around like lemmings (if not zombies), unable to seize the initiative towards a deliberative climate change policy response.

Zombies and lemmings notwithstanding, the organising principles that act as crystallised inhibitors might partly explain why climate change policy has failed us all. Lipari's (2014) concept of interlistening (faithfully or mindfully listening, thinking, and speaking) might hold the key to neutralising these inhibitors and reaching a more plausible conclusion – a valid mental model of the climate change object. Faithful listening and reflective thinking, however, is not something consciously or formally taught in schools or universities by educators (Davies, 2011; Nickerson, Perkins, & Smith, 2014). Furthermore, interlistening is not something that is ordinarily innate or comes naturally through maturity, but something learned (see for example Evans & Frankish, 2009). Similarly, although 'advanced systems thinking' (or critical thinking) is a fundamental reason for higher education, it is perhaps more of an expected outcome rather than a deliberate one. As Martin Davies (2011) recounted:

> There is not a university today (in Australia at least) that does not proudly proclaim that their graduates will – as a result of a degree program in their institution – learn to think critically. Further, there is rarely a subject taught that does not offer the opportunity to acquire skills in critical thinking. However, where is the evidence that we teach critical thinking in higher education? Disturbingly, despite our best intentions, it appears we may be teaching very little of it.
>
> (p. 254)

Is Davies hinting that we are producing more zombies than critical thinkers? Are the mundanities and pressures of modernity giving sustenance to the old mind? If you are treated like a zombie you might as well become one. It is apt here to consider the argument proffered by Ornstein and Ehrlich (2011) who lamented the 'serious and dangerous mismatch' posed by the old mind as it remained entrenched in contemporary thinking and reasoning. Ornstein and Ehrlich (2011) called for an 'educational evolution', declaring that:

> The time has come to take our own evolution into our own hands and create a new evolutionary process, a process of conscious evolution. The human predicament requires a different kind of education and training to detect threats that materialize [*sic*] not in instants but in years and decades – we need to develop 'slow reflexes' to supplement the quick ones. We need to replace our old minds with new ones.
>
> (p. 12)

A natural question is whether it is possible that listening and reflecting – two cardinal principles required for both interlistening and System 2 reasoning – can overcome the automated conventions of the old mind, which is imbued with old ideologies? Ideology has the capacity to infiltrate the conscious, analytic reasoning engaged in System 2 – the ubiquitous and often mindless adoption of 'institutional facts' suggests as much. Moreover, the innate heuristics more common to System 1 are not simply dismissed out of hand when System 2 is engaged. In multiple studies, Stanovich and West (2008) found that some cognitive biases, such as 'belief bias, probability matching rather than maximising, and matching bias' could be overcome by cognitive ability but others could not; such as, 'the conjunction effect, framing effects, anchoring effects, outcome bias, and base-rate neglect' (p. 672). There are, however, different kinds and levels of cognitive ability, and the strength of a bias might vary and be easier or more difficult to overcome. But in the case of unconscious thinking, can a mental model be corrected if it does not know it is wrong? Indeed, why question something that is believed to be true? No further reasoning would be required and the default mental model would prevail.

For instance, natural variability in the Earth's climate system has occurred throughout history, henceforth, why should a changing climate represent anything extraordinary? In the political scheming of things, this deduction represents a palpable counter argument, if one is actually needed, to those advocating greater action to avert a supposed climate crisis. But climate change by all empirical standards is true belief, and collectively, governments presumably have the power to avert such a crisis. Enterprising governments disposed of a neoliberal mental model might enlist 'climate capitalism' as a means of solving the climate change problem. This would, at the same time, bolster their free-market ambitions. Examples of climate capitalism include the commodification of CO_2 emissions and extreme climate events; the privatisation of disaster relief and emergency rescue services; and the introduction of financial instruments such as CAT bonds (catastrophe bonds) used to raise capital in expectation of a disaster caused by climate change.

But in terms of an increasingly polluted atmosphere, the sale of fresh air is perhaps the most enterprising. Sydney entrepreneurs John Dickinson and Theo

Ruygrok now sell fresh Australian air to the Chinese for AU$18.80 a can. Recent studies indicate that air pollution contributes to the deaths of about 1.6 million people in China each year – approximately 4,400 people a day (Rohde & Muller, 2015). I suspect that the commodification of climate change and its associated causes and manifestations are only limited by one's imagination, or the destruction of our planet, whichever comes first.

Political economist Jamie Peck (2010) referred to this 'machine-like rationality of the market' as 'zombie neoliberalism'. The presumption that the privatisation of profits and the socialisation of private debt was dead and buried by virtue of the GFC was premature. The consequence of zombie ideas cannot be killed off so easily even if they have the capacity to bring down financial markets; hence, the inherent danger in a neoliberal governmentality that was not expunged but embraced. This neatly captures the zombie mental model that is obsolete but refuses to accept its obsolescence. Ironically, neoliberal ideologues consider shoring up the financial system by liberating markets and socialising risk as fundamental to human survival. Neo-zombies are administrating our financial system. Meanwhile, the bearers of socioeconomic hardship – exiled to a life of austerity – must endure unaided but also continue to consume. The irony here is superlative. Even though climate capitalism can hardly be seen as a responsible socio-political response to solving the climate change problem, because it fits neatly into a neoliberal logic and is endorsed by powerful politico-elites, it remains neoliberal man's rationale to solving climate change while making a profit. What else could neoliberal man think?

Public choice would suggest that it is not the system that is at fault but the State itself. The State, or more precisely the self-interested cooperative of actors and institutions that constitute the State, can be conceived as a marketplace in which public officials trade power for rents and reallocate resources at their discretion. To resolve the negative consequences of this anomaly, political economist Richard Robinson (2004) proposed that eliminating government intervention through fundamental neoliberal biases, such as the privatisation of public assets and the deregulation of financial markets, would also, *prima facie*, eliminate the problems of rent-seeking[12] and corruption per se. I suspect, however, that this is optimistic. Human ingenuity would find an alternative path, whether semi-legal, illegal, or immoral, to procure the profits that the neoliberal mindset promises.

Further towards a definition of what neoliberalism actually is, Pierre Dardot and Christian Laval (2013) argued that 'Neo-liberalism was the *rationality of contemporary capitalism*' – it is neither an ideology nor economic policy but a social progression propagated on the universal principle of competition (p. 4). The competition to acquire wealth is certainly a social progression of the Highest Order. But it is also primitive and rapacious in its resolve. Perhaps, then, the most apt definition of neoliberalism, however harshly cynical, came from the 2012 movie *Killing Them Softly* directed by Andrew Dominik. Imbued with a neoliberal backdrop and with subtle suggestions of political corruption, hitman Jackie Cogan, played by Brad Pitt, colourfully declares that America is not a country but a business; and as such, strongly reiterated the capitalist credo of payment now due for his services rendered. Perhaps this crude analogy is one not far removed from reality. America, and presumably other capitalistically inclined countries of a similar nature, is no longer a country but a marketplace whose sole purpose is to facilitate the exchange of goods and services for monetary recompense.

Conclusions

Extending on the Institutions Process Model, the Mental Model Process (the Model) provides a more focussed explication of how mental models might form. The mental model process involves a sequence of unconscious and conscious phases and filters that converge and diverge in the quest to extract or construct a personally satisficing mental model of an object, scenario, or phenomenon. A mental tug-of-war might ensue, and contrary to the norm, a disposition for reflection might win out. Alternatively, cognitive limitations might impose the default mental model. The default mental model, brought to mind quickly and intuitively, is more likely to be deficient when faced with matters of probability and complexity.

Probability theory threads its way through this story. It provides a foundation to explain different thinking patterns. Mental model construction is founded on probability theory: the probability of the image coming quickly or easily to mind (the availability heuristic which might lead to biasness); the probability of the image matching the object, scenario, or phenomenon (the pattern matching heuristic which endeavours to match reality); and the probability that the event will occur (the representativeness heuristic which is confounded by the conjunction fallacy). The problem as it stands is that the evidence is always incomplete. Inferences from such evidence must then always be probabilistic.

Humans are forever endeavouring to predict the unpredictable. They are regularly 'fooled by randomness', to borrow the title of Nassim Taleb's 2004 book. We are, in essence, deterministic thinkers: 'Our thinking is too deterministic' (Jones, 2015, p. 25). Humans have a propensity to arrive at directly causal and infallible explanations for phenomena, and thus, in terms of the former, fail to consider all the possible relationships that might have contributed to that causality. Similarly, infallibility rests on the foundations that our explanations will hold true in all scenarios. There are no scenarios in which causal relationships will not hold true. We commonly rely on heuristics to support this worldview. Heuristics are founded on probability. As German theoretical physicist Werner Heisenberg (1958) declared, 'what one deduces from an observation is a probability function, a mathematical expression that combines statements about possibilities or tendencies with statements about our knowledge of facts' (p. 50). As such, one might be left lamenting the probabilities or simply perplexed – contingent on one's grasp of probability theory and his or her depth of reasoning and knowledge.

In recent times, the issue of divergent mental models in relation to the cause and effect of climate change has become a focal point. Faced with extreme climatic conditions, would the layperson recognise climate change as a significant contributor to its cause? Updating one's default mental model to accurately reflect the scientific reality of the day, when humans have a disposition toward observed association as a basis for causal explanation, and the human brain harbours a preference for cognitive economy, is a pervasive challenge. Cognitive economy can be associated with System 1 reasoning. System 1 reasoning is limited in its capacity and not proficient at performing certain tasks that require abstract and/or hypothetical reasoning (Evans, 2014a). In contrast, visionaries such as Leonardo, Galileo, Einstein, and Tesla, appear as highly proficient System 2 mental-modellers.

So what prompts us to think collectively and more deeply? Although Kruglanski's (1989) Lay Epistemic Theory (LET) and 'the fear of invalidity' portends toward lay thinking and as such common sense, which is associated with a System 1 mind mode and generally adequate for the everyday; climate change, for example, is not an everyday problem. Furthermore, in a general sense, evidence suggests that accurate reflection is uncommon in human behaviour. In contrast to the essential cognitive capacity and proclivity associated with those persons disposed of a more conscious endeavour wherein the validity of a conclusion is an imperative. Under ordinary circumstances, the degree of endeavour is positively correlated to what is at risk. But although the risk of dangerous climate change is high, the response remains muted by comparison. This suggests that we have indeed embraced risk, as Beck (2013) proclaimed.

To escape the risks associated with Beck's world, one might claim refuge in the virtual world. Greenfield (2015) has argued that the two-dimensional virtual world is 'so vivid and mesmerising' that it outshines the dull mundanity of the real world. The proliferation of social media platforms such as YouTube, Facebook, and Twitter is perhaps evidence enough of the profound migration to, and urbanisation of, the virtual world. It has been suggested, however, that too much time in the virtual world can lead to Internet Addiction Disorder and the zombification of society – or more precisely, a stressful condition attuned to a mental disorder that substantially impairs an individual's capacity to function normally. Our minds become disorganised. An unconscious enterprise colonises one's system of thought making reflective thinking and reasoning virtually impossible.

In summary, humans have a disposition for System 1 reasoning that is unconscious, intuitive, and fast, or a disposition for System 2 reasoning that is conscious, reflective, and slow. Validating an empirically accurate mental model is more likely associated with the latter (Norman et al., 2017). The former is evidenced by research that suggests that the conscious mind is often elsewhere – leaving the unconscious mind to rule in its absence. As Gifford (2014) explained: 'much of today's behaviour simply follows yesterday's behaviour' (p. 547); which portends to a zombie-like procession through the mundanities of the everyday.

The most familiar neural pathways lead to default mental models that might be accurate or deficient and are quickly retrieved. Less familiar neural pathways lead to more abstract and complex mental models that might be more accurate or less deficient and are slower to retrieve. The latter requires considerably more cognitive exertion than the former. The fundamental goal of which is to validate an accurate mental model of the object, scenario, or phenomenon of focus. Judgement of the probability of an outcome should be anchored to a plausible base rate, which is often not intuitively obvious (see for example Kahneman, 2011). Furthermore, new information (updates) should be assessed in terms of a plausible base rate, and the diagnosticity of the evidence should always be questioned.

In reality, we have a tendency to place much confidence in highly unsound judgements of a probabilistic nature, a phenomenon that Kahneman and Tversky (1973) called 'illusion of validity' (p. 249). For example, an intuitive judgement based on a timeframe of reference (our life experience to this point, for instance) would construe that a dramatic shift in climate is of an extremely low probability (our initial estimate of belief). But our initial estimate of belief is at odds with a plausible base rate because we failed to update our default mental model of the climate change object in light of new evidence. And in the case of climate change, new evidence is constantly coming to light.

It follows, then, if validating a truthful mental model of climate change is contingent on the truth in the premises, or the validity of the cues, then routinely integrating new evidence – true belief – into the model is requisite to success – or more explicitly, to survival. Survival, in instances throughout history wherein the human species failed to think long-term, or did not have the capacity to do so, has proven overwhelmingly disastrous. Suffice to say, neoliberal man is a short-term thinker. Short-termism has become normalised. 'Quarterly capitalism', in which economic performance is measured quarter by quarter, at the expense of long-term policy investments in climate catastrophe prevention, is invariably the govern-mentality and overriding mental model of the early twenty-first century. We think by propagating the present we can postpone the future.

Notes

1 Reflection was 'defined semantically as the consideration of alternative hypotheses (either classifications or problem-solving sequences) when many alternatives are available simultaneously' (Kagan et al., 1964, p. 33). Moreover, in this instance, 'Reflection does not refer to delay that is the result of fear of failure, timidity, or inability to generate a solution' (Kagan et al., 1964, p. 33).

2 From 'What Life Means to Einstein', page 8. *Saturday Evening Post*, October 26, 1929 (Calaprice, 2000).

3 I would add that translating/interpreting Bacon's work (written in the 1600s) into contemporary English takes considerable skill. The passage here was largely an interpretation from *The Advancement of Learning* (1605), pages 146–148.

4 Heraclitus (540 BC–480 BC), from Diogenes Laertius, *Lives of Eminent Philosophers*.

5 Although Hume's choice of the word 'vivacity' (or vivacious) seems somewhat unusual at first, Schmitt (2014) interprets it to mean 'that beliefs are 'more present' to the mind than other ideas, have 'greater firmness and solidity', and the mind is 'more actuated' by them (p. 100). In short, beliefs make a greater impression than do

conceptions. Nisbett and Ross (1980) refer to the same as 'vividness'; the degree of impact or impression incoming information has on individuals. That is, some experiences leave a lifelong impression while others simply pass through.

6 Theodore (Ted) John Kaczynski, the Unabomber, is perhaps the most famous, or infamous, example of such a paradox: Kaczynski is exceptionally intelligent and Harvard educated.

7 'Regression effects' are analogous to 'belief updating'. Belief updating is the process of beginning with some estimate of belief which is then updated or revised as new information becomes available – the logic of how individuals change their mind in light of new evidence (defined as Bayesian Theory). An error occurs when the initial estimate of belief (a representation based on a prior belief or base rate) does not equate with a plausible base rate.

8 For an overview of the science-religion debate see Ruse (2004) and Ruse (2007).

9 Abnormal psychology was defined by Comer (2013) as: 'The scientific study of abnormal behavior [*sic*] in an effort to prescribe, predict, explain, and change abnormal patterns of functioning' (p. 2).

10 Not so coincidently, the U.S. invasion of Haiti was executed under the guise of installing a liberal democracy but was, according to Rhodes (2001), part of an expansionist push aimed at becoming the premier power in the Caribbean.

11 Note that Gilbert Ryle, in his classic work *The Concept of Mind* (1949), strongly denounced Descartes's dualism referring to it as 'the dogma of the Ghost in the Machine' (p. 16).

12 According to Stiglitz (2013), 'rent-seeking' can be defined as the myriad ways in which political processes benefit the rich at the expense of the rest. Including, among other things, government transfers and subsidies to industries and organisations, laws that encourage uncompetitiveness in the marketplace, the failure to properly enforce competition law and/or punish offenders, and statutes that allow corporations to take advantage of others and/or improperly pass their costs onto society.

7 Deductions

De-framing the politico-mediaist

At the first instance, the essence of politico-media framing would construe that the mind is a petri dish for propaganda. I have suggested as much in the previous chapters. Conversely, could it be that framing effects and the propaganda they might purport to instil are overstated and not universally generalisable? In comparatively well-educated societies, which are becoming more informed in today's interconnected world, it is doubtful that the publics are that credulous to be constantly induced by politico-media framing. Presumably, most are exposed to similar politico-media hyperbole day in, day out. There is naturally a choice of news media network and mode. However, much of the news reported is generic. Due to assumptions about acceptance by the largest possible audience, and limitations on the capacity of transmission and storage it is always highly selective. Albeit the news is framed with a slightly different edge contingent on the network's ideological settings and the axiomatic need to make a profit, switching from one news channel to another reveals not much in the way of diversity.

The news media is the politico-media's universal mouthpiece – 'the speak' – but what about 'the listening and thinking' and the role of the decoder and consumer of knowledge? Institutions might create their own collective truths, institutional facts, and formulae, but they are operated and populated by individuals. All individuals are members of institutions to varying degrees, and are therefore more or less institutionalised. Even so, they retain the agency to think and act as autonomous beings. Thus, although the publics *en masse* might be exposed to the same news media, ideologically opposed mental models continue to flourish. For example, and as a general rule, liberals ostensibly support a comparatively constructive approach to climate change policy action; whereas, the approach taken by conservatives could be described as one of indifference. Adherents of each political persuasion for the most part appear to follow suit – a reflection in the main of the ideology of party members.

Political actors have more or less embraced media logic to increase their exposure while accepting some loss of autonomy. Politicians market themselves like any other commodity. Among their most valued assets are spin-doctors and media merchants that frame and reframe public issues in accord with a prescribed political agenda. Assuming, then, that politicians have been mediatised (see for example Meyer & Hinchman, 2002; Strömbäck & Van Aelst, 2013), politics and media are one and the same, the media is 'all powerful', and the publics are broadly susceptible to framing effects, then presumably, individuals would have succumbed *en masse* to the power of the politico-media complex. The common assumption being that the phenomenon of mass media dependency leads to perceptual distortions and a loss of reality for society as a whole (Schulz, 2004).

But if this were true and the media had captured the publics' thinking and acting then the publics would be largely thinking and acting in much the same way. This is, in effect, a reference to psychologist and political scientist Robert P. Abelson's (1964, 2014) theoretical model in which the social networking of interconnected individuals would cause attitudes to gravitate toward the weighted average of opinion and ultimately result in a monoculture. The idea of a monoculture, however, is poorly supported by empirical evidence. I submit that the reason for this, in part, is because the decoding as well as the encoding of a communication requires an act of interpretation. There is a bias in every interpretation and thus every communication. Decoders – referred to herein as validators, consumers, users, filterers of information and knowledge – are prone to confirmation bias. They seek to interpret evidence in a way that is partial to their well-established beliefs, expectations, and theories. From an alternative perspective, if indeed the publics have stopped listening to political speak and gravitated toward a consensus that parallels a monoculture, as described by Abelson, is this illustrative of the zombification of society referred to herein?

According to Richard Maxwell and Toby Miller (2015), the majority of environmental communication scholars would concur that biases and ideologies – about climate change, for example – can be attributed to media campaigns that peddle misinformation, such as those circulated by Rupert Murdoch's Fox News Channel and *The Wall Street Journal*. Contrary to this assertion, extensive research has suggested that the mass media's influence is not nearly as potent as the backgrounds and preconceived worldviews that its readers and listeners hold, which are much stronger determinants than anything delivered in news media content (see for example Entman, 1993; Gamson & Modigliani, 1989; McCroskey & Jenson, 1975; Scheufele, 1999).

Considering the organising principles that individuals harbour, and the capacity of culture to shape a user's mind and his or her environment, it is ostensible that the decoder applies just as many frames and filters as the encoder – the politico-mediaist in this instance. I would go further to contend that it is likely that the decoder applies a greater array of frames and filters. This is because the unconscious mind is normally on autopilot, automatically blocking and filtering incoming data. I might add here that to a growing extent, social media runs a constant stream of interference to the obliviousness of almost everything else. To create meaning, then, requires the interaction of both the encoder and decoder (sender and receiver). This suggests that the impetus of sophisticated media frames to affect a prescribed mental model is somewhat exaggerated, to the extent that the effect is nominal or transient. As found by James N. Druckman and Kjersten R. Nelson (2003), elite framing effects are short-lived and deliberation (or reflection) enhances opinion quality to virtually eliminate its influence.

In theory, and perhaps in reality too, one might surmise that the same people who subscribe to the power of the press are the same people who work in it and tune into it. The assumption being that 'everyone is as engrossed in the news media as we are', which might not be the case. As Jaan Valsiner (2014) pointed out: 'the person who thinks about 'behaviour' is one who at the same time is involved in behaving' (p. 8). Indeed, not everyone tunes into media news. Research by Papathanassopoulos et al. (2013) comprising 11 liberal democracies in the Americas, Europe, Asia, and Australia, found that up to 25 per cent of the respondents surveyed rarely consumed media news, preferring to disconnect from both news media and politics.

The Pew Research Centre (2014) described a similar trend. They found that although one-third of American adults now watched the news online, growth had slowed considerably from 27 per cent between 2007 and 2009 to 9 per cent between 2009 and 2014. This aligns with a survey of 31 European countries by Aalberg, Blekesaune, and Elvestad (2013). They found that while the number of diverse television programs and channels had substantially increased, viewing time had not. The result was a decline in time spent watching programs about current affairs, news, and politics. This supported an earlier survey by Webster, Malthouse, and Ksiazek (2010) that found that the growing abundance of cable television choices and widespread Internet availability had created countless opportunities for consumers to largely avoid the news. Their results indicated that almost half of respondents were 'news avoiders' while the other half were 'news-seekers'.

Finally, a recent report by the mobile analytics company Flurry Analytics, released in early 2017, revealed that by late 2016 American consumers were spending five hours a day (up from about two and a half hours in 2013) on their mobile phones – the ubiquitous artifact of the twenty-first century. Growth was driven by 'communitainment' (communication for the sole purpose of entertainment) and social media was the main app category of use. Social, messaging, media, and entertainment applications accounted for 50 per cent of time-spent. Other categories included gaming which accounted for 11 per cent, and lifestyle and shopping at 5 per cent. News and weather were relegated to near bottom at 3 per cent, above only sport at 2 per cent of time-spent. Suffice to say, news media is being rapidly consumed by social media and was all but out of the competition for time-spent. I suspect that there is profound relief in switching off the news. Moreover, research by Stephanie Edgerly (2015) revealed that those consumers that did actively follow the news were inclined to consume only content that agreed with their ideologies. This had the effect of reinforcing established worldviews rather than developing new ones. If an individual held a particular worldview about climate change, for instance, the news media was unlikely to change that worldview.

It is evident that films and documentaries continue to build a strong case in support of human-induced climate change as opposed to both natural variability in the climate system and climate change denialism. Many of which have featured influential celebrities such as Arnold Schwarzenegger, Matt Damon, Harrison Ford, Leonardo DiCaprio, and Jessica Alba. Media productions such as *An Inconvenient Truth* (Gore & Guggenheim, 2006); *The 11th Hour* (Conners, Petersen, & DiCaprio, 2007); *Climate of Doubt* (Upin & Hockenberry, 2012); and the Emmy-award-winning documentary *Years of Living Dangerously* (Bach & Gelber, 2014), have raised awareness of the dangers of climate change and environmental degradation. This, combined with a rapidly declining news media audience and a decline in public opinion with respect to the institution of journalism that has paralleled public institutions (Carlson, 2017), makes it difficult to establish a significant causal link between politico-media framing and climate change inaction. Moreover, it poses the likelihood that one's belief or disbelief in climate change can be attributed to something more profound than politico-media framing.

Reinforcing this assertion are studies that reveal politicians, chief executives, and government institutions are disliked and distrusted the world over. As far back as the early 1970s, American political scientist Arthur H. Miller (1974), after studying a series of surveys conducted in the 1960s, remarked on the publics growing political distrust in the ruling elite. He declared this mistrust as not something transient, but a long-lasting and entrenched form of political cynicism. Suffice to say that Miller was correct in his assessment. Fast-forward to the 2000s and distrust in political leaders and their institutions is endemic (Fiske & Durante, 2014; van Prooijen & van Lange, 2014; Zeineddine & Pratto, 2014). A study by the Pew Research Center (2017) revealed that trust in the United States government has been in a downward spiral since its initial study in 1958, and was now at historic lows. In 2017 only 18 per cent of Americans trusted their government to do what was right, as opposed to about 75 per cent in 1958. As a consequence, individuals are more likely to deliberate and seek advice from a more trusted source before accepting the views of politico-elites, irrespective of how well they frame an issue: 'Our collective good sense tells us not to trust them [politico-elites] until proven not guilty' (Fiske & Durante, 2014, p. 103). As Linda L. Carli (2017) pointed out, in terms of our decision-making, opinions, and behaviours, 'expert and likeable' sources are more influential than 'inexpert and unlikeable' sources. And as this investigation suggests, it is easy to make the assertion that politicians are in the main, not likeable and not climate experts.

This raises the critical issue of what is asked of society in a liberal democracy. If citizens of voting age are forced to vote, as is the case in many countries such as Greece, wherein voting is mandatory, and Australia by way of a fine imposed for failing to do so, it can only be construed as a moral dilemma to advocate for the collective mental models of government that are, for the most part, believed by that Society to be untrustworthy; and/or dislikeable and incompetent as the case might be. Forced to vote for the lesser of two evils is hardly a liberal democracy.

Studies that purport to the significance of framing effects might garner more attention than those that do not. Druckman (2004) argued that scholars mainly focussed on those studies that resulted in statistically significant framing effects and that these studies were overrepresented in journals because of a bias toward positive findings. Supporting this was a systematic review by Chiu, Grundy, and Bero (2017) that found that 'spin', a biased representation referred to as 'science hype', was prevalent in biomedical literature. Spin, in scientific literature, is designed to distort the interpretation of results and mislead readers with the intent that results would be viewed more favourably. It is evident that the pressure to publish and the pervasive influence of the consumer marketing machine has infiltrated scientific literature, to varying degrees, much to its detriment as a trusted source of empirical evidence.

Extending this, Druckman and Nelson (2003) concluded that the impact of framing effects on an audience was conditional. Controlled experiments failed to account for the interventions of interpersonal conversation and deliberation that take place in the everyday which had a moderating and reasoning effect on frames. Similarly, individual thinking and acting is subject to groupthink and peer pressure – 'what would my friends and family think'? or 'what would they do in this situation'? For example, Druckman (2001a) noted that endeavours to replicate Kahneman and Tversky's (1984) famous framing experiment (the Asian Disease Problem) have met with mixed results with some failing to achieve significance and others matching the findings of the original experiment. Philosopher of Science Paul K. Feyerabend (1981) has suggested that all scientific theory must be constructed so that it is 'testable', and immediately discarded if the test does not conform to the predicted outcome. In light of this, are academic journals susceptible to mediatisation? Do we cling to popular theorems (and experiments and methodologies) merely because they appeal to our senses rather than conform to empirical evidence? Hence the 'replication crisis' in science. Is Kahneman and Tversky's Asian Disease experiment a case in point? Are the effects of framing on public opinion exaggerated? (see for example Bechtel, Hainmueller, Hangartner, & Hebling, 2015; Cacciatore, Scheufele, & Iyengar, 2015).

I suspect that endeavours to simulate the idiosyncratic nature of politico-media framing can achieve only a roughly biased evocation of reality as a *bona fide* milieu would be virtually impossible to replicate. For example, the widely acknowledged disconnect between self-reported behaviour and observed behaviour. It is not uncommon for an attitudinal response, expressed in the discourse of the simulated world, an interview, for example, to contradict actual behaviour. Similarly, with respect to the observer effect, applied in a sociological or psychological sense, in which the act of observing alters the observed. For example, the timing, wording, and choices posed in an opinion poll influence the way participants think and respond. Prominent planning theorist, John Friedmann (2011), argued that in the social world, which is the real world, the luxury of controlled conditions is not possible: 'one has to proceed by inserting theories into the flow of life'; then by observation, continually revise both theories and actions based on actual results (p. 30).

The question is, then, if a particular ideology is firmly implanted in System 1 reasoning can framing have any mutable effect at all? Does framing have the capacity to infiltrate one's instinctive trust in intuitive belief? Leaving religion aside, which is a special case, evidence does suggest that knowledge and therefore education – level of education is the single best predictor of climate change awareness – with respect to climate change scepticism or denialism, does have the propensity to increase acceptance that climate change is occurring, it is human-induced, and is a substantial threat to all life on Earth (Guy, Kashima, Walker, & O'Neill, 2014; Lee, Markowitz, Howe, Ko, & Leiserowitz, 2015). The framing of climate change has much less a propensity to achieve the opposite.

According to Thuraisingham (2013), 'Knowledge is key, as we are not consciously aware of the frames of reference, rules of thumb and preferences that are hardwired into our brains' (p. 24). The inference being that knowledge has the capacity to circumvent this hardwiring. Knowledge has the potential to avert culturally and emotionally driven judgements that lead to poor decision-making. Mindful that knowledge here is true belief and not information masquerading as knowledge.

In summary, the influence of sophisticated framing effects adopted by the politico-media is overstated and a declining influence with respect to the publics' thinking and acting. I emphasise three main points. First, the historical intervention of Internet technology has created a virtual world that is capable of providing more information and variety than the real world. Or at least the Internet makes information and interconnection exceedingly more accessible and thus it is simple to escape the banality of politico-media speak for something more personally enriching and culturally specific. Corporate actors still play a role, however, a stream of multifarious rivals competing for consumers' time, money, and attention have diminished their dominance. For liberal democracies, citizens have the opportunity now more than ever to participate in the democratic process. But equally so, there is the option to 'tune out' to the real world and 'tune in' to the virtual one as a means of escapism and relief from the drudgery of the everyday. I would add that the mobile phone (or now more aptly, the smartphone) is now the American consumer's mobile communication device of choice. In 2016 Americans spent about five hours a day on their mobile phones but only 3 per cent of time-spent was dedicated to news media – nine minutes on average out of a possible 300. The antiquated mass media, inhabited and operated by the politico-media, is now a distant memory. The 'new media' is narrower and richer and more diverse, and personalised in its content.

Second, the mental models of individuals, in general, have inherited certain idiosyncratic organising principles that operate more or less independent of the conscious mind. Consequently, incoming information and knowledge might be unconsciously rejected and thus denied, even though the politico-media goes to great lengths to encode its messages in a suitably biased frame. The politico-media's reach might be limitless, in principle, but even the most sophisticated framing techniques have limits and constraints as decoders consciously and unconsciously institute their own gatekeeping mechanisms. As argued by Nicholas Carah and Eric Louw (2015), powerful elites use the media to disseminate their preferred meanings – and endeavour to engender a 'false consciousness' – but this does not guarantee that their preferred meanings will have the desired effect. As a media enterprise, meaning-making is at best, 'messy and opportunistic' (p. 16).

Third, I cite a longitudinal online survey of the Australian publics (N=17,493), conducted 2010–2014 by Leviston, Greenhill, and Walker (2015), that is especially germane to this inquiry. In its most recent year (n=5163), respondents who believed that climate change was human-induced pointed to 'scientific research' as the primary source for their information. Moreover, university scientists and environmental scientists were the most trusted source for truthful information about climate change. The government, car companies, and oil companies were the least trusted. Those respondents who believed that climate change was a result of natural variability or was not occurring were more likely to point to common sense or historical weather patterns as the basis for their reasoning. Notably, 'very few' respondents cited information from politicians, government, or the news media as the basis for their reasoning. Indeed, the results of this survey suggested that political speak had lost the trust of the publics and rendered its political leaders as largely irrelevant. Suffice to say, source credibility is a quintessential precursor to persuasion. Quite simply, people are more likely to yield to the influence of people they trust and like (Carli, 2017; Petty & Cacioppo, 1986). In accord with the points foregoing, the evidence to indicate that the duplicitous framing of climate change by the politico-media was the prime suspect in the failure of efficacious climate change policy, was relatively weak.

A general cognitive proclivity

In light of the theoretical models introduced in Chapters 4, 5, and 6, specifically, the recurring intuitive and reflective juxtapositions of the mental-modelling process, whether at an institutional or individual actor level, I surmise that the failure of climate change policy is associated with a general proclivity to engage System 1 reasoning and remain in that mind mode. I refer to this as 'a general cognitive proclivity' based on the concept of the cognitive miser model (Fiske & Taylor, 2013; Taylor, 1981) and dual-process theory (Evans, 2003, 2007, 2010, 2014b). *General* meaning the majority of humankind; *cognitive* as it relates to the thinking and reasoning processes and capacities of the human mind/brain; and *proclivity* meaning an innate and adaptive human disposition to conserve mind/brain energy by intuitively engaging a System 1 mind mode.

Moreover, the disposition for a System 2 mind mode might be evolutionary and correlated to intelligence. For instance, Stanovich (1999) argued an association between System 2 reasoning and general intelligence. He has suggested that System 1 reasoning is independent of general intelligence. Relevant here is that an individual's capacity to form a truthful mental model of the world is a function of his or her level of consciousness and intelligence. And as Innes, Gilchrist, Friedman, & Tompkins (2016) put it when deliberating on Dewey (discussed in Chapter 1): 'Intelligence is behavior (*sic*) (including both thinking and doing) that leads to more effective adaptation' (p. 33). Based on the premise that only through adaptation can one survive.

In summary, I propose that some individuals have a strong inclination to remain in a System 1 mind mode and a similar disinclination to engage System 2. This is in contrast to individuals that have an equally strong disposition for reasoning in a System 2 mind mode. System 1 is characterised by shallow, more intuitive thinking and reasoning as the brain fights reflexively to conserve energy. System 2 is characterised by deeper, more reflective thinking and reasoning which requires the brain to expend comparatively more energy.

It is apt to point out that a justification to conserve brain energy might enlist one or more of the following: (1) objective facts are impossible to obtain; (2) the particular knowledge sought does not exist or; (3) the knowledge is purposively avoided because it causes anxiety. With respect to the latter, social psychologist Leon Festinger's (1962) theory of cognitive dissonance suggests that the human organism strives to achieve an internal consistency, a 'peace of mind' of sorts, among opinions, attitudes, values, and knowledge. There is a presumed consistency between what an individual knows and how he or she acts. Festinger called this consistency 'consonance'. In contrast, an inconsistency between what an individual knows and how he or she acts might result in psychological discomfort. Festinger called this inconsistency 'dissonance'. In the struggle to reduce dissonance, individuals will endeavour to actively avoid the objects, scenarios, and knowledge that might increase it. In this sense, if how one acts, for example one's penchant for a beachside way of life, is inconsistent with the knowledge that sea level rise and intensifying storminess, as a manifestation of climate change, is a serious threat to beachside living, then one is motivated to avoid expending cognitive energy to acquire such knowledge and modify his or her mental model accordingly. Adopting a general cognitive proclivity, then, serves to diminish cognitive dissonance.

The Deductive Model: the Ping-Pong Effect

The human endeavour to comprehend the workings of the brain that produces the mind is a formidable challenge. In technical terms, the brain's grey matter is composed of billions of neurons that are interconnected by a network of dendrites. Electromagnetism governs the electrical signals that race through the neural pathways like search engines. Confounding this machine-like structure is a ghost in the machine: consciousness, 'the properties of which far exceed the sum of internal connections and electrochemical discharges that give rise to it' (DeSalle & Tattersall, 2012, p. ix). Consciousness is a mysterious thing. It disappears during sleep and reappears upon awakening. Beyond that, a genuine scientific understanding of consciousness is not apparent. For the purpose herein, consciousness is that 'thing' associated with the mind/brain that distinguishes an enlightened and attentive state from that of a zombie state. One might only contend that a higher level of consciousness is associated with System 2 thinking and reasoning.

Dual-process theory suggests that the mind/brain has two capacities: intuitive and reflective – or System 1 and System 2. The mental model theory fits neatly into dual-process theory (discussed in Chapter 3). Evans (2003) made an obvious but commonly overlooked observation: we cannot underestimate the importance of the distinctively human faculty that System 2 endows. Without engaging System 2 we cannot hope to learn by experience in order to avoid, for example, the perils of nuclear war and unconstrained global warming. A superior level of thinking and reasoning are the mechanisms that have allowed *Homo sapiens* to flourish and continually adapt to their environment. However, the diminution of the dangers of climate change, the dumbing down of the climate change debate, and the infantilisation of democracy, suggests that the evolutionary-recent System 2 has succumbed to a more primitive form of cognition – System 1 – that is shared by both humans and animals. As Johnson-Laird et al. (2015) explained:

> the neglect of falsity in mental models leads to systematic fallacies, counterexamples occur spontaneously in the evaluation of inferences as invalid, and deductions depending only on mental models (system 1) are easier than those depending on a fully explicit model (system 2).
>
> (p. 11)

In the case of System 1, then, reasoning processes that are inherent and purport to heuristics that evolved to solve specific problems of adaptation that confronted *Homo sapiens* extemporaneously, are ubiquitously engaged. The danger in this is that errors in judgement and decision-making are more likely and, in particular, prediction, and thus the advantages of planning in advance, is less likely. In essence, System 1 is reactive rather than proactive. The focus is on the 'here and now' rather than the state of affairs 20 years from now. To achieve a concerted global climate change action, one must transform the 'here and now' into a matter of urgency, which requires the capacity and proclivity to think and act beyond one's own lifetime. As Beck (2010) opined:

> Why is there no storming of the Bastille because of the environmental destruction threatening mankind, why no Red October of ecology? Why have the most pressing issues of our time – climate change and ecological crisis – not been met with the same enthusiasm, energy, optimism, ideals and forward-looking democratic spirit as the past tragedies of poverty, tyranny and war?
>
> (p. 254)

Why is not one thunderstruck by the significance of the climate change problem, as Beck exclaimed? Has apathy been normalised? What triggers the switch from apathy to urgency? Perhaps, in response to Beck's (2010) critique,

Hugh C. Willmott (2014) pointed to 'science as a privileged form of knowledge in modernity' and the collective failure of scientists to diligently disseminate the likely build-up of greenhouse gases in the atmosphere when it first became evident (p. 22). Considering the early climate-related work by Tyndall (1861) and Callendar (1938), the possibility of catastrophe only became broadly acknowledged in the 1980s with the discovery of a hole in the ozone layer. Also bearing in mind that the technology to extract fossil fuels from the Earth, the burning of which is the main culprit of climate change causation, was made possible by science. Despite these uncompromising truths, we hold to an obstinate bias for limitless growth that defies the logic of a finite planet.

Although it is evident that both logical and belief-based reasoning are concomitantly engaged, a conscious endeavour to adhere to a logical imperative will reduce, but not eliminate, inhibitors such as matching and belief biases. Crystallised inhibitors that facilitate deficient mental models, such as a misunderstanding of probability theory, embodied cognition, and a range of other cognitive biases can, therefore, be moderated by consciously engaging System 2. However, this is by no means a natural response to complex ecological and social problems. As studies by Houdé et al. (2000) found, the jump from System 1 to System 2 (a change in the subject's reasoning strategy) requires the mind to inhibit a perceptual process in order to activate a logical reasoning process which requires a shift in the cortical anatomy of reasoning from the posterior part of the brain to a left prefrontal network. This brain shift is requisite to accessing logical thinking. The prefrontal cortex, the foremost part of the frontal lobe, is where rational thought is largely processed.

I surmise that the reasoning process attributed to System 2 cannot be purely syllogistic. Evans (2003, 2007) purported to this by noting that the two minds in one brain (System 1 and System 2) compete with one other for 'control over our inferences, decisions, and actions' (p. 458). They exist in a state of mutuality and conflict. The end product is therefore a by-product or composite of both Systems 1 and 2. By extension, rather than simply a valid deduction as a consequence of the analytical processes that occur in System 2, the final mental model product comprises artefacts from both Systems 1 and 2. And although mental model validation, post exploring all the available possibilities, is presumed to occur in System 2, it would be unlikely that validation would survive wholly unadulterated by the substantial competition brought to bear by System 1. Moreover, Mike Oaksford and Nick Chater (2001) have suggested that most reasoning is based in System 1, and System 2 is used more or less to validate the conclusions that System 1 generates. I suspect, however, that the process is not so one-dimensional.

Li-fang Zhang (2013) proposed a construct of 'mode of thinking' (traditionally left/right brain dominance or hemispheric thinking) as '*analytic* (originally left brain dominance), *holistic* (originally right brain dominance), and *integrative* (originally whole-brain)' (p. 31). I propose that the mental model process might resemble something more attune to that illustrated in Figure 7.1. This model, extrapolated from the theoretical model presented in Chapter 6 and broadly construed from work on dual-process theory by Evans (2003, 2007, 2010, 2014b), implies an integrated matrix – similar in effect to Ron Sun and Xi Zhang's (2004) Clarion Model that described an integrated dual representational structure that unified implicit and explicit knowledge. The Model exhibits an adaptable thinking and reasoning process (multiple systems that interlock to perform different functions) as the search for a valid mental model jumps back and forth from intuition to reflection like a ping-pong ball. As described in Kahneman's (2011) depiction of wisdom as a product of the interaction between fast and slow thinking that conflate to suppress the biases inherent in cognition. This suggests that individuals think in two minds simultaneously, although greater emphasis or preference is concentrated in either one at a given time in response to different goals: the generation of possibilities derived from the imagination or true belief, as opposed to the validation of those possibilities, for example.

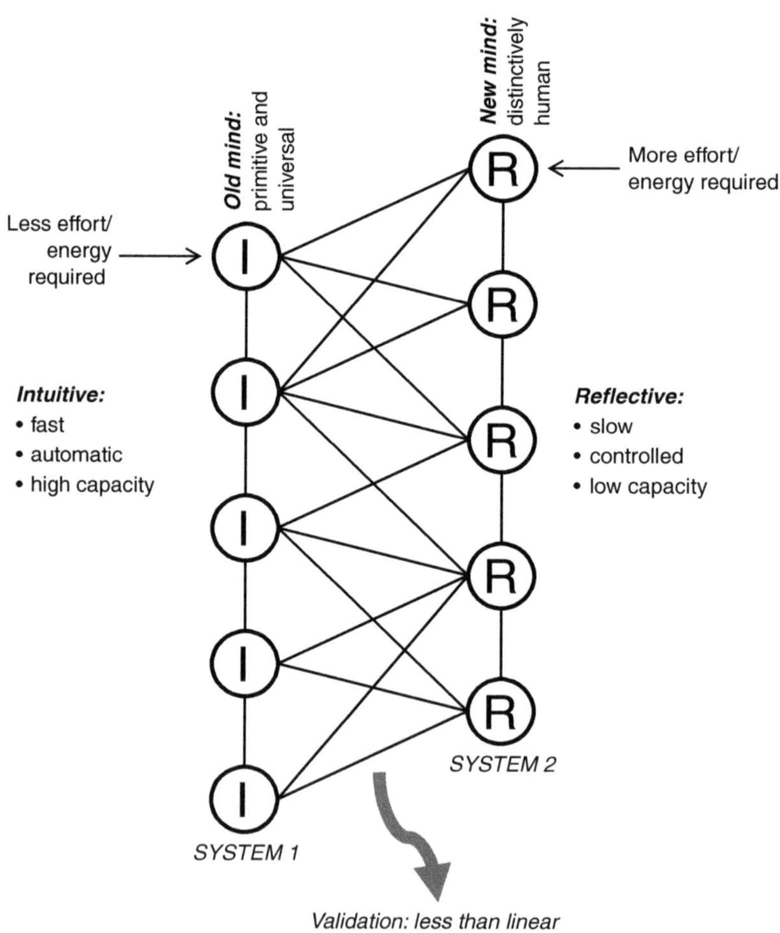

Figure 7.1 In Two Minds and the Ping-Pong Effect.

The Ping-Pong Model depicts the largely unconscious battle that rages between the intuitive mind and the reflective mind (I=Intuitive, R=Reflective). As Thuraisingham (2013) argued: 'Any intellectual achievement is always a mixture of both – a combination of intuition and logical deductive processes' (p. 8). That is, intuition and cognitive biases might be common to System 1 but cannot be wholly divorced from System 2. An individual's biases, or ideological settings, are largely irresistible: the irrational element of creative intuition cannot be suppressed (Popper, 1959). The challenge is to correct for errors of bias in intuitive judgement and retain what is valid and useful. Achieving validation might involve an immense number of neural firings. Even the simplest of thoughts might generate more than 100 neural firings (or pings; Kaku, 2014). I suspect that there is an automated drive to synchronise Systems 1 and 2 to achieve some semblance of a 'peace of mind', to avoid the cognitive dissonance as described by Festinger (1962). 'Peace of mind', however, does not motivate one to take deliberative climate change action, and is more likely to result in the opposite.

The 'ping-pong effect' more aptly describes 'natural logic', or perhaps intuitive logic as previously discussed. The ping arbitrarily bounces back and forth between intuition and reflection. Where the ping stops, however, might have closer ties to genetics than it does to a learned reasoning aptitude. For example, though some errors in deductive reasoning can be attributed to belief bias and other unknown factors, Melanie Stollstorff (2014) has theorised that errors can also be attributed to emotion as a product of genotype: 'the serotonin transporter gene, known to influence attention towards or away from emotionally valenced information, can affect logical reasoning and the brain mechanisms that support this process' (p. 15). Although 20 per cent of humans carry this genotype there is no evidence to suggest that it cannot be overcome. I suspect that genetics play an important role in triggering mechanisms that either promote or prohibit logical thinking and deductive reasoning.

Because System 2 reasoning is grounded in deductive reasoning (analytical or hypothetical thinking) it is less likely to construct or validate a deficient mental model. Deductive reasoning entails sorting through an array of possible mental models that might require modification, or constructing new ones until a useful possibility is found that is analogous to the user's interpretation of the incoming data. Learning to inhibit (or perhaps recognise) the cognitive biases that are more common, but not exclusive, to System 1 is key to effectively engaging System 2. But engaging System 2 is not a guarantee of success. Despite an increasing capacity for reasoning that comes with age, some biases also increase with age. Furthermore, as previously discussed, a disposition for cognitive economy (the cognitive miser) suggests that engaging System 2 is more effortful and thus less desirable. There is a natural proclivity to expend less energy on powering the brain and, as a consequence, follow the most familiar neural pathway to a default mental model without questioning its validity. This might be referred to as 'narrow mindedness'.

For *Homo sapiens* to survive a perilous world, the human brain evolved to process information in the fastest and most efficient way possible. It was simply a matter of the quick or the dead. As *Homo sapiens* evolved, the concentration of pings moved progressively toward R in the Model (reflective thinking and reasoning common to System 2). System 2, the 'new mind', allowed its user to solve novel problems through reasoning about the future. Whereas the 'old mind' of System 1 relied primarily on experiential learning and evolved cognitive systems wherein behaviour was repeated because of past success. From this it is evident that the old mind is incapable of solving the novel and slow-moving problem of climate change. Climate change is critically a new mind problem. Presently, however, we are endeavouring to solve a new mind problem with old mind thinking.

System 1 reasoning is limited by 'association' – the fundamental propensity to rely on observed association as a foundation for causal explanation and the basis for learning in most animals (Jones, 2015; Kahneman, 2011). The user is fatefully drawn into repeating the same actions because of past success, even though the problem has transformed into something different, and perhaps more threatening (the infallibility bias). This is known as the 'Einstellung Effect', and it is at the heart of the problem-solving conundrum.

The Einstellung Effect occurs when a solution to a problem comes immediately to mind because it has worked previously in a similar context. This pattern of 'fast thinking' precludes any alternatives from being considered and constitutes the downside of previous experience. For instance, it has been shown that politicians and scientists can be so entrenched in their preferred theories that they fail to objectively evaluate valid evidence that is contradictory (Gould, 1996; Tetlock, 2005). There is thus the possibility that climate change denialists have succumbed to the Einstellung Effect. This would also mean that they have failed to recognise the problem of climate change as something novel and dangerous, resorting to an outmoded recourse rather than a newly defined commensurate response.

Thuraisingham (2013) explained that automated fast thinking is premised on association. As most thinking is associational there is a propensity to think about climate change in the same way – that causality is more intuitively attributed to natural variability by association. Associations to climate change are not nearly as perceptible. Our 'weekend-focal' perceives the transformation from sunshine to rain as simply associated with a change in the weather. A drought remains a drought even though it has not rained for decades. Any closer inspection requires more abstract and probabilistic reasoning. According to Robert G. Jones (2015), both the disposition and capacity to take deliberative action is related to mental ability. Individuals with greater mental ability are typically more altruistic; therefore, mental ability is associated with taking action for the common good. But precisely what 'mental ability' constitutes is not clear.

It is broadly accepted that intelligence involves the human capacity to learn from experience and adapt to one's surroundings (Sternberg, 2017). From the mid-1980s this definition was expanded to include metacognition, how individuals understand and control their thinking processes. More recently, the role of culture was included: different cultures have different concepts of what it means to be intelligent. Robert J. Sternberg (2017) thus summarised intelligence this way: 'intelligence is the capacity to learn from experience, using metacognitive processes to enhance learning, and the ability to adapt to the surrounding environment; it may require different adaptations within different social and cultural contexts' (p. 145). As discussed in Chapter 6 (System 1 and System 2 reasoning), being conscious of one's thinking processes, metacognition, is most relevant herein.

Theories of intelligence are plentiful. Extensive research has centred on a 'general mental ability' or 'general intelligence' (simply termed '*g*') as the best predictor of performance across an array of mental applications (Kaplan & Saccuzzo, 2012; Spearman, 1904, [1927] 2005). Although a consensus has not been reached in terms of its hierarchal structure and composition, Charles E. Spearman's ([1927] 2005) general factor ('*g*') of intelligence remains prominent in psychology. Linda S. Gottfredson (1997) defined intelligence as 'a very general mental capability that involves the ability to reason, plan, solve problems, think abstractly, comprehend complex ideas, learn quickly and learn from experience' (p. 13). It is not simply book learning but reflects a broader and deeper comprehension of one's environment and a capacity for sense-making and working out what action to take. For the most part, it is the ability to cope with cognitive complexity: the more complex the task the greater advantage *g* provides. This suggests that '*g*' is predominantly attuned to a System 2 mind mode. Consequently, some minds are simply better adapted to solving particularly complex problems, such as climate change, than are others.

According to Lev Vygotsky and Alexander Luria (2013), animal instincts are the prototype of emotions and traces of the predatory animal's fight and flight instincts can be found in human fear and anger. This implies that displays of fear and anger are largely associated with the 'old mind' and System 1 reasoning. An individual prone to such displays might be classified as a predominantly 'System 1 thinker'. Ingolfur Blühdorn (2007) might have been referring to System 1 thinkers when he argued that 'late modern societies', despite their vociferous criticism of what he called 'merely symbolic politics' and their declarations to take decisive action, have neither the will nor the capacity to 'get serious' about environmental dangers such as climate change. Although System 2 reasoning is obligatory for survival in the contemporary world, modern man has retreated to a pervasive pattern of endemic System 1 complacency. Indeed, an air of complacency has washed over the world. The human resolve to confront global problems has been reduced to zombie status – and a Dark Age of human thought.

Human thought, and the capacity to think, has humble beginnings. John L. Bradshaw (2014) surmised that the brain's evolutionary path was probably episodic, occurring in phases – static for the most part then interrupted with short bursts of rapid change. Although the human lineage remains unresolved, a common theory contends that about 4.2–3.8 million years ago a cooling climate caused the forests to recede forcing arboreal-climbing hominids from the trees and onto the open plains. This culminated in an evolutionary transformation to bipedalism. Thus, the early phase of human evolution likely originated as a consequence of climate change. Climate change forced early humans to make the jump from the trees to walk the Earth. This prompts the question: in the current phase of human evolution, which is confronted by an array of global perils such as climate change, what would force the jump to System 2 reasoning and a higher level of consciousness? Theoretically, the negative consequences associated with the dilemma of a dangerous climatic change would presumably motivate individuals to think more deeply about a solution.

A general negativity bias

Hibbing, Smith, and Alford (2014) proposed that, with respect to political orientation, 'of the many differences between liberals and conservatives is the nature of their physiological [of the body] and psychological [of the mind] responses to features of the environment that are negative', which is referred to as a general negativity bias (p. 297). Hibbing et al. argued that conservatives are more perceptive to negative stimuli than are their political counterparts, the liberals. In short, conservatives are more sensitive to environmental stimuli and thus react more strongly to signs of danger than do liberals. Furthermore, Hibbing et al. argued that conservatives are less open to new ideas and are therefore less likely to elicit novel information that might turn out to be harmful. Similarly, they are more likely to retain positive information and refrain from taking chances with the unknown.

At the first instance the response to the threat of climate change would appear to contradict the hypothesis posited by Hibbing, Smith, and Alford (2014) that conservatives are more sensitive to negative environmental stimuli than are liberals. Assuming that this hypothesis holds weight: 'individuals who are physiologically and psychologically responsive to negative stimuli will tend to endorse public policies that minimize [*sic*] tangible threats by giving prominence to past, traditional solutions' (p. 304). Hibbing, Smith, and Alford (2014) make reference to 'tangible threats' which infers an important distinction between the brute corporeality of war, and the comparative abstract incorporeality of climate change, for instance. This suggests that conservatives respond pragmatically to tangible problems but less so to intangible ones like climate change. The theoretical worldview known as the 'rigidity of the right' implies the same. Republicans, for example, as opposed to Democrats, are more likely to ignore or deny the established science on climate change (Mazur, 2018).

Feelings of fear and uncertainty compel conservatives toward order and structure in society, and thus a desire to maintain the status quo. A study by Schreiber et al. (2013) suggested as much. They found that liberals and conservatives engaged different cognitive styles – to the extent of possessing 'significantly different' brain structures – when assessing risk and uncertainty. Conservatives' brains were hardwired to respond with more intense physical reactions to threatening stimuli but only when the image presented was abhorrent and explicit. This hardwiring extended to cognitive processes and policy preferences. Schreiber et al. also conjectured that acting within the coordinates of a partisan environment might alter the actor's brain to overrule the attributes that heredity ascribed.

That said, the theory that a general negativity bias exists – inherited through our genes – which is associated with an individual's political persuasion has been proposed in recent times (see for example Hibbing, Smith, Peterson, & Feher 2014; Lau, 1982, 1985; Schwartz, 2014; Weaver, 2011). For instance, a study by Nicholas Smith and Anthony Leiserowitz (2014) found that discrete emotions, particularly worry, interest, and hope, were strongly associated with increased support for climate change policy. With worry being the strongest predictor of advocacy for climate change policy. Moreover, an impact study by Leiserowitz (2004) consisting of a survey of American adults who watched the disaster movie *The Day After Tomorrow* found that the climate change scenario depicted in the movie altered the attitudes of viewers. Viewers were found to have a heightened level of concern about climate change after watching the movie to the point where it shifted their conceptual models (mental models) of the climate system and even influenced their voting preferences. In a broader sense, Valsiner (2014) endorsed this theory by explaining that humans are not content to simply experience a bad experience, they have to reflect on it to give it meaning. Simply put, the human mind/brain is disposed of a negativity bias which means that it has greater recall of, and response to, negative experiences and information than it does to positive experiences and information.

Baumeister, Bratslavsky, Finkenauer, and Vohs (2001) declared that 'bad is stronger than good', and 'it may be highly adaptive to respond more strongly to bad than good' (pp. 323, 362). Substantial evidence indicates that a general negativity bias exists that might indeed be evolutionary and encourage certain thinking and acting. I submit the theoretical proposition that negativity bias is the mechanism that provides the motivation for the broad recognition of a problem (System 1 reasoning and a System 1 response) and the need for a sound solution (System 2 reasoning and a System 2 response). Attaining a coherent problem definition, comprehensive decision analysis, and innovative solution and plausible deduction, however, is contingent on an individual's general cognitive proclivity. This is the fundamental and generally unconscious disposition to conserve energy and avoid confronting adverse and/or complex problems and paradoxes by constricting thinking to a System 1 mind mode, as it competes with the user's capacity and proclivity for System 2 reasoning as a function of the ability to truthfully and creatively mental-model the future.

To summarise, although there is a general need for humans to be optimistic (Jones, 2015), there is also a general disposition, across both conservatives and liberals and the publics in general, to a negativity bias. It is postulated here that negativity bias fills the gap as to what stimulates the cognitive miser and compels him or her to engage in more effortful thinking and reasoning. In this way, the motivation to engage System 2 is driven by the desire to diminish one's negative feelings with regard to the object of focus (negative stimuli). It is presumed that this is achieved by validating a plausible conclusion.

Conclusions

Popular media theory has it that the public agenda is orchestrated by the news media which ultimately shapes public opinion. In contrast to this, I conclude that humanity's response to climate change, for example, is more about reasoning and reflection than it is about politico-media 'sleight of hand' and sophisticated framing effects. That is, whether one advocates strong climate change action or simply abstains is conditional on one's capacity and proclivity for deeper, more reflective thinking and reasoning rather than politico-media framing. More precisely, a person's orientation to a System 1 or System 2 mind mode and the mental models formed as a consequence is predictive of his or her thinking and acting about climate change. Moreover, I submit the theoretical proposition that the motivation to jump from a System 1 mind mode to a System 2 mind mode is associated with a general negativity bias.

I also conclude that the failure of climate change policy is associated with a general proclivity to engage System 1 reasoning and remain in that mind mode – a zombie state, so to speak. I refer to this as a general cognitive proclivity. The intuitive mind mode of System 1 uses less energy than the reflective mind mode of System 2. The brain is a high energy-user and has a natural desire to conserve energy (Magistretti, 2015). Furthermore, a disposition for a System 2 mind mode, which is the exception rather than the rule, might be an evolutionary cognitive adaptation and correlated to intelligence. For instance, Stanovich (2009) has suggested that System 1 reasoning is independent of general intelligence. One could fairly argue that this fits the description of the mindless zombie – primitively pursuing its most basic instincts of survival.

Relevant is that an individual's capacity to form an accurate, or truthful, mental model of the world is conditional on his or her capacity and proclivity for System 2 thinking and reasoning and therefore, his or her level of intelligence and consciousness. In this way, System 2 is the alter ego of System 1: the mindful human being as opposed to the mindless human zombie. To elucidate this further I constructed a model (Figure 7.1) that described a 'ping-pong effect' wherein thinking and reasoning jumps back and forth between System 1 and System 2. This model was based mainly on the work of Jonathan Evans (2003, 2007), a cognitive psychologist, who developed the 'two minds in one brain theory' (System 1/System 2). According to this theory, two minds in one brain compete for ascendency over the user's thinking and acting. This state of mutuality and conflict results in a conflation of both Systems 1 and 2. So, rather than simply a consequence of the analytical processes of System 2, the mental model constructed is a conflation of artefacts from both Systems 1 and 2. That is, in the endeavour to validate an accurate mental model of climate change, for instance, the mind/brain's 'cognitive search engine' jumps back and forth from intuition to reflection like a ping-pong ball. Validation, however, remains the providence of System 2.

I have also proposed that negativity bias is associated with a disposition for System 2 thinking and reasoning but this disposition is subject to, among other things, an individual's level of consciousness. That is, negativity bias is an antecedent to making the jump from System 1 to System 2. But in its simplest capacity, negativity bias reflects 'the tendency to perceive danger rather than reward related information' (Williams et al., 2009, p. 804). Thus, it is a response to a negative that might well yield the avoidance of an outcome bias – indicative of a departure from rational behaviour – but only if its user has a capacity and proclivity for System 2 reasoning.

Extensive research confirms an association between genes and negativity bias (see for example Ashare et al., 2013; Lesch & Canli, 2007; Pornpattananangkul, Cheon, & Chiao, 2014), indicating that a negativity bias is inherent and an evolutionary cognitive adaptation: a response to the comparatively greater consequence of threat – in the endeavour to survive – as opposed to reward. In effect, humans are hardwired to take more notice of negative stimuli as opposed to positive stimuli. Studies have shown that the human brain recalls and responds to negative experiences and information by up to three times that of positive experiences and information.

Negativity bias can be defined as 'the propensity to attend to, learn from, and use negative information far more than positive information' (Vaish, Grossman, & Woodward, 2008, p. 383). An example of this was noted by Ulrich Beck (2015): he observed that a traumatic experience, such as Hurricane Katrina,[1] induced a process of reflection in which connections between things were made that were not thought of before. Hurricane Katrina brought to bear the issue of racial inequality, for instance. In her aftermath, racial and class diversity emerged as a polemic issue. It was purported that racial discrimination, in the case of African Americans, was the reason emergency services were so slow to respond and disaster relief was impeded. In this way, the Hurricane Katrina experience brought attention to other negative aspects of society. From this, a general negativity bias can be understood as a primary motivator that humans have developed to avoid or provide a solution to obvious and not so obvious problems. Or more broadly, temper the fears and frustrations of the boundless mysteries of the universe: the fear that we are all alone in the universe. For instance, the Nobel Prize winning physicist Richard P. Feynman (1992) developed a set of mental models while at high school that later helped him solve the considerable conundrums posed by quantum mechanics.

It is important to emphasise here that a negativity bias does not subvert imagination but enhances it. The English poet John Keats (2002) expressed this eloquently as 'Negative Capability':

> several things dovetailed in my mind, and at once it struck me, what quality went to form a Man of Achievement, especially in Literature and which Shakespeare possessed so enormously – I mean *Negative Capability*, that is when man is capable of being in uncertainties, Mysteries, doubts, without any irritable reaching after fact and reason.
>
> (p. 60)

Similarly, the British author and academic, Richard Holmes (2008), in his book *The Age of Wonder: How the Romantic Generation Discovered the Beauty and Terror of Science* quoted Humphrey Davy, one of the forefathers of modern Chemistry writing in 1807: 'Imagination, as well as the reason, is necessary to perfection in the philosophic mind. A rapidity of combination, a power of perceiving analogies, and of comparing them by facts, is the creative source of discovery' (p. 276). Indeed, the imagination is free to run wild before being captured by the reality of reason.

In simple terms, the problem of climate change and its well-documented and publicised negative connotations activates one's negativity bias and stimulates inquiry – or ought to – which thus triggers the jump from System 1 to System 2. But this is conditional on one's capacity and proclivity for System 2 thinking and reasoning. A natural question, then, is do you think about climate change in a System 1 or a System 2 mind mode? What mental images do the words 'climate change' bring to mind? Or do any images come to mind? Zhang (2013) argued that 'intellectual styles'[2] might change automatically as a function of socialisation, but, if individuals want to purposively develop a specific intellectual style they require a consciousness of their current mind mode and the determination to acquire a new one. In essence, humans require a high enough level of consciousness to ascertain whether they are reasoning in System 2 or, habitually disposed to System 1 and are thus, in poor control of their cognitive faculties. Furthermore, research has indicated that an individual's DNA holds the answer as to why some individuals are more predisposed to logical reasoning (common to a System 2 mind mode) than others (Stollstorf, 2014; Stollstorff, Bean, Anderson, Devaney, & Vaidya, 2013a; Stollstorff et al., 2013b). In light of this, it is in the DNA of System 2 thinkers, whether policymakers, politicians, or publics, to rise to the challenge of climate change.

Notes

1 Hurricane Katrina slammed into the city of New Orleans in August 2005 – taking the lives of 1800 people and causing damage estimated at US$125 billion (these figures vary markedly).

2 As defined by Zhang (2013): 'Intellectual styles – an umbrella term for such constructs as cognitive styles, learning styles, teaching styles, and thinking styles – refer to people's preferred ways of processing information and dealing with tasks' (p. ix).

8 Summary and conclusions

The multifarious manifestations of climate change are many and mystifying. Even if one has perchance met one of its manifestations in the flesh, and recognised it as such, it remains difficult to accurately mental-model. This can be explained, in part at least, by embodied cognition, which posits that the mind works in tandem with the physical world (Lobel, 2014). Although much conjecture surrounds embodied cognition, Robert D. Rupert (2009) defined his hypothesis of extended cognition, a component of 4E Cognition (see Aizawa, 2017) this way: 'human cognition literally includes elements beyond the boundary of the human organism … and the human mind and self spread into the environment beyond the organism' (p. 241). In effect, there is a cognitive system that comprises the brain, body, and environment. Think, metaphorically, of the 2009 movie *Avatar* as the ultimate version of extended cognition. The Na'vi, an alien species living on the planet Pandora, a lush tropical paradise bestowed with both beauty and danger, are intimately connected to their environment through an intricate network of neural tendrils. In essence, the Na'vi and their environment assimilate to form one global consciousness.

As extended cognition implies, if there is a disconnect between the brain, body, and environment, something as esoteric as climate change is not so easily embodied in one's consciousness. Scientific uncertainty adds to this, providing sceptics, denialists, and other ideologues with a false consciousness as to its true identity. In addition, climate change the object, expressed in politico-scientific language, might be misconstrued or difficult for the publics to clearly visualise. Science illiteracy and insufficient knowledge of probability theory, compounded by the often emotive and long-standing unintelligible liberal/conservative political debate/disarray that dogs deliberative climate change policy – the persistence of which is a redundant enterprise and in no-one's interest – leaves a gap that is necessarily filled with 'something'. Should a deficient mental model of climate change ensue, it has the propensity to diminish advocacy for efficacious climate change policy. From this worldview, the lack of a definitive and consistent climate change narrative might create the impression that the climate change problem is not a priority.

For instance, coastal dwellers are reluctant to accept the climate change policy of a planned retreat. Evidence indicates indifference toward sea level rise, coastal erosion, and storm surge inundation, by both government authorities and beachside residents. Proposals to withdraw from vulnerable coastal zones are frequently met with anger and suspicion. They are viewed as authoritarian because they impinge on private property rights. Although faced with the prospect of intensifying storminess, seaside property owners are reluctant to modify their way of life. Furthermore, citizens are commonly determined to return to the scene of devastation, as in the case of New Orleans and Hurricane Katrina. Although this mainly relates to necessity and popularism, both government and citizens are constrained in their thinking and reasoning by a deficient mental model of the climate change threat and the profound danger it represents.

Pivotal to my story was climate change policy failure. Critics will argue that examples of climate change policy success are apparent. But these are isolated. My argument here focusses on global success, as this is the only focus that will avert dangerous climate change. The sticking points are considerable. The cost of climate change prevention is a contentious issue. Debate continues over the appropriateness of cost–benefit analysis, specifically the discount rate used, as a methodology to value climate change policy investments. Humans have an inherent desire to value the present over the future – consumption today as opposed to the glacial pace of sea level rise that might pose a threat some 20 years from now. The exigencies of the moment trump long-term thinking. This inherent peculiarity strongly favours investing in today. This renders costly investment in preventive measures as counterintuitive, and provides impetus to the now familiar narrative that there are other more pressing issues to contend with. In short, much is tied to the cost of climate change prevention and climate change policy investments – who should pay and how much, and thus the concept of sovereignty and the capitalist ideal.

A deficient mental model of the climate change object would subsequently lead to a deficient policy – inadequate design and implementation or a failure to implement – and therefore policy failure. In essence, if the mind cannot build a concrete bridge between the physical world and the climate change object, a gap results that is filled with unsubstantiated belief, unrealistic optimism, superstition, and/or opinion. Alternatively, the gap might simply be filled with perplexity: a directionless parable that makes no sense. The climate change narrative, then, might become distorted and rejected as alarmist fiction, or even a hoax. Although becoming less common today, this type of narrative has undermined advocacy for efficacious climate change policy. In summary, endeavours to institute a coherent and consistent climate change narrative remain mired in politico-media speak. Consequently, as pointed out by Hulme (2013), there can be no clear climate change narrative, only a selection of narratives each framed with its own emphasises – and thus de-emphasises.

The questions

The aim of my story was to answer three questions. I hope to have addressed each one adequately: (1) To what extent is the politico-media framing of climate change associated with the failure of climate change policy? (2) To what extent is the mental-modelling of climate change associated with the failure of climate change policy? (3) If there are relationships within and/or between (1) and (2) above, how do they combine to either aid or impede the successful design and implementation of climate change policy?

For the first question: 'To what extent is the politico-media framing of climate change associated with the failure of climate change policy?' I concluded that the capacity of politico-media framing to modify public opinion, with respect to climate change policy per se, is overstated. Although an 'association' probably exists to a limited extent, the strength of this association could not be corroborated.

For the second question: 'To what extent is the mental-modelling of climate change associated with the failure of climate change policy?' I concluded that there was substantive evidence to suggest that the capacity and proclivity to engage System 2 reasoning has a strong bearing on the human capacity to recognise climate change as a genuine threat and respond in kind. This is because, based on a general cognitive proclivity, some individuals have an inclination to remain in a System 1 mind mode (intuitive) and a similar disinclination to engage System 2 (reflective). System 2 reasoning is a distinctively human facility that allows users to avoid disasters, such as climate change, by accurately mental-modelling the future. As climate change is future-oriented, this would suggest that inferior mental-modelling, that is prone to a System 1 mind mode, is associated with the failure of climate change policy per se.

For the third question: 'If there are relationships within and/or between (1) and (2) above, how do they combine to either aid or impede the successful design and implementation of climate change policy?' I concluded that a relationship exists between mental model aptitude and the failure of climate change policy. The more proficient one is at mental-modelling the future the more likely it will be that one will take action to prevent climate change. Moreover, a tangible relationship between the politico-media framing of climate change and the failure of climate change policy was not corroborated.

The investigation

Two main causes of climate change policy failure were identified and investigated: the politico-media framing of climate change (nurture) and the mental models formed and held of climate change (nature). I proposed that the politico-media framing of climate change was associated with the formation of disparate mental models of climate change; which in turn, was associated with the failure of climate change policy. I concluded that although the politico-media framing of climate change might be associated with climate change policy failure, it was overstated and not a prime motivator in mental model formation. Underpinning this conclusion was the difficulty in establishing a link between framing effects and its capacity to modify public opinion; and the fact that the decoder, as well as the encoder, was required to create meaning in a communication. In addition to the strong case made in film and documentary that advocates human-induced climate change and thus contradicts the denialist narrative.

There was also evidence to suggest a general distrust in politicians, based on the half-truths of political communications, rational choice theory, and public choice theory, that might prompt an audience to seek further confirmation about the claims politicians and the politico-media make. Also questionable was whether the news media commanded the undivided attention of the masses. In a world of mass media digital devices, the evidence indicated that the publics' consumption of news media was rapidly shrinking. As de Botton (2014) put it: 'A flourishing life requires a capacity to recognize [*sic*] the times when the news no longer has anything original or important to teach us' (p. 255).

From the historical worldview of nature versus nurture, I concluded that politico-media framing was a social evolution construct and a form of modern-day nurture, or even an increasingly ineffective endeavour at social engineering. It was evident however, that natural evolution was the dominant impetus associated with climate change policy failure – as well as its success. The formation of disparate mental models – as a consequence of one's capacity and proclivity for System 2 reasoning – was the likely suspect complicit in the failure of climate change policy. The formation of a mental model that accurately depicted climate change was the defining feature of those actors and institutions that advocated urgent climate change action. Mindful that mental models are blueprints for thinking and acting in a certain way.

Mental-modelling, then, was a consequence of natural evolution – the ongoing evolution of human consciousness. In terms of climate change policy failure, nature (in reference to the capacity and proclivity to engage a System 2 mind mode and accurately mental-model climate change) wins out over nurture (in reference to the politico-media framing of climate change). I therefore arrived at the conclusion that the failure of climate change policy is associated with a general proclivity to engage, and remain in, a System 1 mind mode. There is a preference to retain this mode as the rudimentary means to negotiating the everyday. I described the prevalence of this mode of thinking, as it extended beyond the everyday to the constant, as the zombification of society. The zombie mode is intuitively primitive. It is based on the singular premise of 'find and consume'. Of course, there are those that are more conscious than unconscious, and thus remain among the living, so to speak.

I surmised that a general negativity bias – the disposition to 'pay attention to, reflect on, learn from and use', negative information far more than positive information, triggered the jump from System 1 to System 2. I emphasise that it is a general negativity bias that initiates the jump to a System 2 mind mode, and not the negative stimuli or the emotion, such as fear, that might or might not be attached to it. The impetus that sparks a negativity bias and the jump to a System 2 mind mode might be disharmony, antagonism, or a contest of wills about whose hypotheses are correct; or the sudden advent of countervailing norms that rupture one's routine thinking and acting. Future survival, as opposed to immediate survival, is perhaps the definitive motivation to engage System 2. Importantly, a disposition for System 2 reasoning as indicative of a higher level of consciousness. And a higher level of consciousness is needed to accurately mental-model the future, and thus the consequence of climate change.

Proficient mental-modelling is reliant on the analytical architecture of the mind/brain which orchestrates System 1 and System 2 reasoning. This requires a certain cognitive choreography to optimise results. Moreover, System 2 reasoning has the capacity to diminish cognitive biases which increases the probability of validating a mental model that has a plausible conclusion. Conversely, a disposition for System 1 reasoning is unlikely to result in a distinction between a valid mental model and an invalid one, unless the reasoner is coerced into doing so.

To conclude, climate change policy fails, whether by design or implementation, for the same reason it succeeds: the human capacity and proclivity – *dunamis* (potentiality) as opposed to *energeia* (actuality) – for a System 2 mind mode which has a greater probability of validating an accurate mental model of the climate change problem; and as such, acts as a blueprint for a commensurate action. Accepting this conclusion meant that the proposition – the failure of climate change policy is associated with the politico-media framing of climate change and the disparate mental models formed of climate change as a consequence – could not be corroborated to any reasonable extent.

The reckoning

Policymaking is the process of guiding society away from bad outcomes and toward good ones. Predicting outcomes as a consequence of policy made is a difficult proposition. In this instance, a good policy outcome is one that prevents dangerous climate change, which is in the interest of both current and future generations. There is a propensity for policy to be captured, however, by the minds of vested interests. That is, the minds of the regulators are captured by the actors and institutions they regulate. This is a difficult problem to counter because the cognitive biases that motivate favourable treatment toward a particular industry or individual are not always apparent to the user, and might be real or imagined by the observer, as both are under the influence of their own set of cognitive biases and default mental models. Mindful that throughout history, broadly speaking, political corruption has been the norm and not the exception.

Corruption aside, which is an unfortunate fact of institutional life, a principle policymaking recommendation is to adopt hypothetical thinking, which is characteristic of a System 2 mind mode. Moreover, in the design and implementation of climate change policy, the default mental model should be recognised for what it is: one outcome of several possibilities that might be deficient in what is known to be true. This means treating each iteration of a phenomenon, a category 5 hurricane for example, as different to all previous iterations of the same phenomenon – the inverse of the Einstellung Effect. And subsequently, develop systems and people to embrace what are potentially novel and perhaps radical methodologies.

A better understanding of how the human brain works, and how mental models are formed, would help to achieve a global acceptance of uniform climate change policy. For example, the acceptance of a carbon tax might be more amenable if policymakers better understood what mental models the different publics held of climate change. The same would apply to policymakers and bureaucrats, and the mental models they hold of climate change. Moreover, if we truly understand climate change – let us presume that we do – the need for a carbon tax now means we failed in the first instance to recognise it. To once again turn a blind eye is tempting fate and thus tantamount to fatalism. What is more, governments have a duty of care to protect their citizens, and the citizens of planet Earth, from dangerous climate change, and should be held personally accountable for failing to do so.

Planet Earth is indubitably warming, and this warming is gathering momentum. Most important is that the base-rate now requires recalibrating and the corresponding mental model updating. This means that oft-used guidelines such as '1-in-a-100-year flood' (a 1 per cent probability of a specified level of flooding to be equalled or exceeded) are made obsolete. For instance, in a world of accelerated climate change, extreme climate events would likely occur every few months (Steffen, Hunter, & Hughes, 2014). By extension, probabilistic thinking and reasoning based solely on historical events no longer apply. Climate change places probability in the realm of what is conceivable – or in terms of mental-modelling, what is imaginable.[1] It is not enough to predict that an extreme climate event will occur, but what that event will look and feel like. In the case of policy and practice, the measure of an individual's deductive knowledge is now more important than the measure of historical experience – which is counterintuitive.

In order to ameliorate the accuracy of mental models formed, dominant thinking and analyses that are institutionalised under oppressive systems, and the politicising of knowledge translation and transmission, need reform. For example, geopolitical pressure to modify the wording in CO_2 emissions reduction agreements and IPCC reports to suit the priorities of individual countries severely compromises the integrity of the stated intent (to diminish and/or prevent the threat of dangerous climate change). The collective goal is overshadowed by the goal of the individual. This equates to 'tribal economics' in a global economy which promotes symbolism rather than substantive climate change action. Perception creates reality until it is wiped out by an extreme climate event.

Despite perceptions that might suggest otherwise, in truth, negotiations to reduce greenhouse gas emissions have thus far failed, due largely to a fragmented global climate governance, institutional complexity, and a profit maxim that defies the logic of long-term thinking. This is demonstrated by the fact that after 20 years of COP meetings and 24 years of repeated warnings by the IPCC, the Earth edges closer to ecological collapse. Given this, the future of planet Earth cannot be left to the discretionary powers of governments. The Earth's wellbeing should not be contingent on climate change policy that shifts back and forth between liberal and conservative ideologues. It is the civil arm of democracy that must succeed in solving the climate change problem where governments have failed.

The perpetrators of pollution must be held accountable. Albeit a complex proposition, lack of accountability cannot be the trade-off for economic growth. It is presently argued that human-induced climate change and ecological degradation be classified a 'green crime' (Hall, 2015; Hall & Farrall, 2013; Kramer, 2014). Climate change, as a consequence of atmospheric pollution, should be treated as criminogenic (Hall, 2015). Rob White (2011) has argued that in retrospect, climate change inaction in the present day will be looked upon as the 'gravest of transnational environmental crimes'. And Beck (2013): 'there is no greater threat to the Western way and quality of life than the combination of climate change, environmental destruction, dwindling energy and water resources and the wars they could spark' (p. 65).

To address environmental crime, Higgins, Short, and South (2013) proposed a law of 'ecocide' that was globally recognised. International lawyer and barrister Polly Higgins (2013), in a submission to the United Nations Law Commission in 2010, defined ecocide as 'the extensive damage to, destruction of or loss of ecosystem(s) of a given territory, whether by human agency or by other causes, to such an extent that peaceful enjoyment by the inhabitants of that territory has been severely diminished' (p. 257). Within this definition two classes of ecocide were further defined: human-caused ecocide and naturally occurring ecocide. Ecocide that was naturally occurring, whether attributable to climate change or not, was the responsibility of governments; ecocide that was human-caused was the responsibility of both governments and business. The aim was to establish a framework for the prevention and prohibition of environmental crime, and also prosecute offenders.

Importantly, a law of ecocide would impose a legal 'duty of care' that all countries were subject to (Higgins et al., 2013). Ecocide would be treated as a crime under the remit of the International Criminal Court.[2] The governments of all countries would be required to act in advance of ecosystem collapse, whether as a consequence of rising sea levels or other extreme climate events such as storm surge. Human-caused ecocide would be treated as 'a strict crime of liability', thus closing the loophole of intent. Whether by intent or by accident, corporate ecocide would be categorised an environmental crime and the perpetrators prosecuted accordingly. Administrators of the design and implementation of policy, policy investments, and development approvals, would be held legally liable for safeguarding ecosystem services and bringing to an end business practices that caused environmental degradation.

I have implied that politicians, because of the important task they are encumbered with, be measured in some way to ensure their suitability and competence to perform such a demanding role. The work of neuroscientists has suggested that the brain architecture of some individuals might have a greater disposition for System 2 reasoning. To this end, I contend that the ability to form an accurate mental model of the future is characteristic of an adaptive visionary. A 'true visionary', I surmise however, is a superior mental-modeller disposed of a goal to enrich the common good. As Kaku (2014) put it: 'Human consciousness, I believe, is the process of continually forming a model of the world, in order to simulate the future and carry out a goal' (p. 268). In the case of our political leaders, even if one's mental model is accurate and his or her vision is clear, it is acknowledged that having the conviction and charisma to achieve that goal in the presence of disbelieving publics is another matter, as Nicola Tesla and company discovered.

The need for people to be System 2 reasoners and thinkers with an aim to accurately mental-modelling the future has critical implications for democracy and the majority vote. If both the publics and the policymakers are unable to grasp the science, then the design and implementation of efficacious climate change policy are unlikely to succeed. Considering the consequences, such failure could prove catastrophic. Acknowledged also is that in a liberal democracy, all people are viewed as equal and deserving of the same opportunities to make a contribution to governing the country. To suggest that particular individuals should be elected on the basis of 'cognitive design' – because they are classed by neuroscientists as supreme mental-modellers – is perhaps the stuff of science fiction? Nonetheless, to place democracy above meritocracy is counterintuitive.

Moreover, it is quite possible that by manipulating a handful of genes one can be made more intelligent, which leaves the door ajar. Alternatively, in a global context, if the collective mental model of reason does not emerge in the case of climate change policy, then we are doomed to disagree and suffer the revenge of 'Gaia', as described by Lovelock (2007a). I subsequently conclude that the relevance of political actors and politics, political systems and governance, or whether the school of thought subscribes to a top-down or bottom-up policy-making approach, is subordinate to the mental models that individuals, groups, and the publics form and hold. Correcting deficient mental models of the climate change problem is therefore a primary course of action.

Sooner or later there must be a reckoning. According to analysis by NASA, the mean global surface temperature of the Earth in 2017 was the second warmest year since records began in 1880 (National Aeronautics and Space Administration, 2018). If the effects of the recent El Niño and La Niña weather patterns were statistically removed from the analysis, 2017 would be the warmest year on record. By comparison, analysis by scientists at NOAA (National Oceanic and Atmospheric Association) concluded that 2017 was the third warmest year on record. The discrepancy in rankings a result of the different methods used by the two agencies. According to scientists at NASA's Goddard Institute for Space Studies (GISS), the global average temperature in 2017 was 0.90°C (1.62°F) warmer than the 1951–1980 mean. Despite colder than average temperatures in some parts of the world, and a levelling off of global CO_2 emissions over the three years to 2017 (Figueres et al., 2017). Moreover, it is important to note that this trend was broken throughout 2017, as Global CO_2 emissions began rising once again (Global Carbon Project, 2017).

A recent report from London's Royal Society (2017) reads as follows: analyses of sea-floor sediments have revealed that the current concentration of

CO_2 emissions in the atmosphere, at just over 400 parts per million (ppm), has not been experienced for at least three million years. In addition, average per annum concentrations of methane in the atmosphere increased from 1800 parts per billion (ppb) in 2011, exceeded 1840 ppb in 2016, and will possibly breach 1850 ppb in 2017. Methane (CH_4) is short-lived compared to CO_2 lasting only one or two decades, but it is an exceptional heat-trapper and thus an exceptional climate changer. Overall, analyses by NASA and NOAA confirm the Earth's long-term warming trend, and more rapid warming over the last 40 years. So, as acclaimed writer Barbara Kingsolver (2015) warned, 'we can either shift away from a carbon-based economy or find a new place to live' (p. 341). The problem with the latter is, of course, as Kenneth E. Boulding ([1966] 2013) duly noted, this is 'Spaceship Earth' and there is nowhere else to go.

It is evident that a more sophisticated understanding of mental-modelling as a precursor to human behavioural change is needed: one that goes beyond accurately mental-modelling climate change to include that critical nexus between humankind, the natural environment, the economy, and planet Earth herself. Moreover, the crux of my story focussed on why climate change policy fails. The natural evolution of this story would suggest focussing on why it might succeed. It is therefore only fitting to close with some advice from a true visionary. In his *Principles for the Development of a Complete Mind*, Leonardo da Vinci advised the following: 'Study the science of art. Study the art of science. Develop your senses – especially learn how to see. Realise that everything connects to everything else' (as cited in Borden, 2017, p. 6).

The end-state

'Justice is the first virtue of social institutions, as truth is for systems of thought' (Rawls, 1971, p. 3). In accord with Rawls's theory of distributive justice, the end-state is one in which the 'distribution of the benefits and burdens of social cooperation' is just, and thus fair. Any inequalities that do arise must be rectified if they cannot be justified. Global climate change amplifies inequality. A technology-driven economic agenda is driving environmental policy and draining the resolve for social justice and equality. The burden of global climate change, which will be largely borne by the poor and marginalised, will greatly amplify this. The poor and marginalised also enjoy little benefit from their social cooperation. A neoliberal state of affairs serves the interests of the politico-elite at the expense of the disadvantaged, literally. This does not augur well with Rawls's theory of distributive justice.

With respect to 'systems of thought', truth can be found in both Systems 1 and 2 but can only be validated in System 2. System 1 is an intuitive system of thought. It is of the old mind and not conducive to solving contemporary problems like climate change and the devastation of the natural world. Even as planet Earth begins to fray at the edges there remains a beleaguered endeavour to govern climate change – in the neoliberal tradition – to institutionalise and commoditise it in accord with the tenets of capitalism. Under this socio-economic model, according to sociologist Marcus Taylor (2015), climate change mitigation has been an 'unqualified failure'. Indeed, climate change represents a profound challenge to the prevailing neoconservative mental model of limitless growth and material use. This school of thought extends well beyond the Earth's carrying capacity and a viable vision of humanity's future prosperity on planet Earth.

I have framed the 'year on year' economic growth model as a zombie enterprise, propagated by capitalist conscripts. It is neither the liberator nor the saviour of society. Our liberal democracy needs rethinking. The price of admission to any liberal democratic system is the Social Contract. All citizens are free and equal but subject to the laws made and enforced by a sovereign power. In modern democracies, the sovereign power takes the form of representative government, which is subject to 'the general will' of the people. But as Rousseau ([1762] 2003) lamented: 'Man is born free and everywhere he is in chains' (p. 1). Although Rousseau wrote this in 1762, it is arguably more relevant today than it was then. The inequality between the rich and the impoverished has reached obscene levels. It grows daily as austerity governance becomes the norm. Quite simply, for the majority of people, living a normal life on planet Earth has become unaffordable. Access to adequate healthcare, education, housing, food, and clothing, is reserved for the privileged few.

There are more slaves in the world today than at any other time in human history. In 2016 'The Global Slavery Index' estimated that there were 45.8 million modern slaves in the world today. Despite the fact that almost all countries have laws prohibiting slavery. Slavery, along with rising inequality, gender-based oppression, institutionalised sexual abuse, global exploitation of labour and resources, and the destruction of our natural world, is in breach of the Social Contract – to be sure. Mere politicians are incapable of solving these problems simply because they are instrumental in them. Indeed, men are not angels and power corrupts. Or to be politically correct, men are not angels and neither are women; as either might succumb to the corruption that power affords. In reality, it is naïve to believe that politicians are especially prudent decision-makers, considering the rules and criteria for selection. At the behest of a politician we can invade another country, persecute minority groups, censor free speech, and spy on everyone.

Human mastery over nature is an illusion: nature tames human; human tames nature; nature tames human, and the cycle continues. One suspects that nature will have the last word, however, as the sun will eventually fade. But equally so, as Hulme (2011) recounted, it would be remiss to succumb to 'reducing the future to climate', as this seriously undervalues humanity's imagination and ingenuity. Nonetheless, in the desperate race to secure, produce, and consume the Earth's resources, and the potential consequences of extreme climatic change, questions abound as to what the end-state might look like. One would hope that the end-state is a higher state of consciousness – a return to the state of nature but without the savagery.

The future state

> to the extent that any utopia advocates change, it is not advocating the imposition of a theory or set of abstract principles onto existing society, but instead the imposition of a "speaking picture" of instantiated principles onto society.
>
> (Barbara Goodwin, *The Philosophy of Utopia* (2001, p. 14))

Political philosopher, Barbara Goodwin (2001), in her endeavour to reconcile the relationship between existing society and the imagined utopian one, pointed out that we already have a 'speaking picture' [mental model if you will] of what utopia is. Of course, on the road to utopia, we might end up somewhere very different from the one we pictured, as many state-sponsored disasters of the twentieth century were the results of grandiose utopian visions gone wrong (Scott, 1998). Some of them in the dystopian image of George Orwell's *Nineteen Eighty-Four* (1949). But, considering the reward, is the road to utopia a risk worth taking? Whether or not we believe it is, for better or for worse, most of us have embraced it through the Social Contract. Perhaps the archetypal definition of embracing risk, in the context of Ulrich Beck's influential work, *World at Risk* (2013), was placing the future state of humankind in the hands of the politico-elite. One might otherwise seek utopia in the promise of the technological singularity, the virtual world, or in one's religious faith. All have bearing to the future as it unfolds. In the meantime, how does one go about rectifying climate change policy failure?

Critical self-reflection as a precursor to critical action is a good place to start. To be sure, an examination of the mental model one holds of climate change, and its manifestations, is requisite to responding with a commensurate action. I suspect that this entails an in-depth meeting with one's own mind. It is evident that the power of reflection lies therein. One might facilitate this meeting by first overcoming group tribalism and the human affliction to acutely categorise others. It is elementary that thinking beyond the majority, and in particular, beyond one's ideological, ethnic, political, and nationalistic tribalism, is fundamental to forming an intellectually valid, or truthful picturing (as would be the unpretentious endeavour) of reality. In addition to this, a revitalised awareness (mindfulness) of the present, while maintaining a deep respect for the past, is imperative to rejecting social zombiism and accurately anticipating, and planning for, the future state of our planet.

I propose that future research be aimed at identifying individuals that have a natural disposition for System 2 reasoning. I also suspect that individuals who think in pictures (or images) have a greater capacity to accurately mental-model the future than those who think predominantly in words. Not to detract from the written word, however, as a language of images requires a language of words in order to manipulate and shape its meaning. Without the literate mind, composing and expressing one's thoughts, as with the story herein, becomes a difficult proposition indeed.

Recognised is that 'governing by cognitive selection' on the basis of mental-modelling aptitude, as opposed to democratic election, is not such a desirable proposition. However, this knowledge might help to better appreciate and understand deficient mental models and go about correcting them in a sensible and open-minded way. This could be accomplished through image-based learning and the use of concept mapping. Personal meaning mapping (PMM), for example, would provide a culturally specific conceptualisation of a given object, scenario, or phenomenon (see for example Whitley, 2013). This could then be compared to an empirical model of the same with an aim to reconcile conflicting mental models of the climate change problem that different actors and institutions and sectors of the publics hold. This could be further extended to include the world's diversity of cultures to mobilise collective advocacy for climate change action.

The links between a general negativity bias and a disposition for a System 2 mind mode, and a System 2 mind mode and a disposition for mental-modelling, require further investigation. The question remains: is mental-modelling something that could be practiced and perfected – perhaps even taught at an early age. Could the school curriculum include Mental-Modelling 101? For instance, as previously mentioned, brilliant theoretical physicist Richard P. Feynman (1918–1988) developed a 'scheme' in high school (the now defunct Far Rockaway High School in New York City) to solve mathematical theorems that he used throughout his adult life (Feynman, 1992). He taught himself to construct a comprehensive set of examples (probable mental models) based on the conditions of a given theorem as they were sequentially defined. Feynman used this 'one box of tools' repeatedly to solve the considerable complexities posed by quantum field theories. It enabled him to validate as true if the theorem matched one in his set of mental models, or false if a counterexample (contradictory mental model) could be found. As Feynman explained, his different method of doing integrals (which he applied to the integral formulation of quantum mechanics) often proved successful at solving complex mathematical theorems wherein orthodox methods failed. Feynman was a self-taught advanced mental-modeller and received the Nobel Prize for Physics in 1965 for his work on quantum electrodynamics (QED).

By extension, although comparatively few people have the cognitive capacity of Richard P. Feynman, is mental-modelling simply an innate phase of the mind/brain's natural evolution and if so, how might this be integrated into computer modelling to unlock the secrets of our past, present, and future? Contrary to this, is the integration of mental-modelling into computer modelling the exact wrong path to take because it would inhibit the natural evolution of human consciousness? Hence the philosophical debate, and to some extent the fear, that artificial intelligence will replace humankind's unique cognitive capabilities.

Recent initiatives such as The Good Judgement Project and aggregative contingent estimation (ACE), which seeks to 'dramatically enhance the accuracy, precision, and timeliness of intelligence forecasts' through probabilistic judgements, mathematical aggregation of individual judgements, and expertise and cognitive style (The Good Judgement Project, 2015, para. 1) are nascent ventures into developing sophisticated mental-modelling techniques. Although these programs are not, on the face of it, synchronous with the evolution of complex scientific phenomena such as climate change, they might indeed be humankind's intrinsic adaptation to these phenomena – through natural evolution – unbeknown to him in a formal sense. I submit, however, that one should not undervalue the power of intuitive thinking nor the imagination. I suspect that one has to imagine to reason and reason to imagine. That is, the picturability of the world depends on one's capacity to imagine; whereas, its intelligibility depends on one's capacity to reason.

With the advent of the Internet – imbued with far greater social and moral responsibility and accountability going forward – might we better develop our understanding of complex phenomena, such as climate change, through images? Image-based learning might be both the beginning and the end of the learning process. We recognise the image of our parents long before any words are spoken. As we grow older, images become central to adaptive mental model elaboration in the performance of specific tasks (Schnotz & Wagner, 2018). Text provides the initial comprehension, but the comprehensive study of images provides the depth of knowledge to respond to the demands of the task. Contrary to this are the pitfalls of a hyper-digital culture reduced to a conglomeration of automations and algorithms, and the zombification of society. The advent of mobile digital technology, for instance, has the potential to erode the human capacity to reflect. Reflection is the key to validation. Validation brings us closer to reality. By changing someone's reality we change the future.

Notes

1 It is not apparent that overcoming the exceptional uncertainties in climate science, to more precisely express the probabilities of future climate change, is imminent. The literature corroborates this. Fuzzy linguistic narratives and constructs used to express socioeconomic and technical developments, and the highly contested probabilistic estimates of climate sensitivity, exist today as they did when Fu et al. (2005) discussed them in their paper 'Beyond Probability: New Methods for Representing Uncertainty in Projections of Future Climate' (see for example the probability range in recent IPCC reports).

2 See Polly Higgins's website: http://eradicatingecocide.com/polly/.

Glossary

Carbon emissions and greenhouse gas emissions terms commonly used in conjunction with climate change. Depending on the source, 'carbon emissions' or more correctly, 'carbon dioxide emissions' (CO_2 emissions), 'greenhouse gas emissions' (GHGs which consist primarily of water vapour (H_2O), carbon dioxide (CO_2), nitrous oxide (N_2O), methane (CH_4), and ozone (O_3), and 'emissions' (CO_2 emissions or GHGs) are often used interchangeably to convey a similar meaning. More precisely, as a measure of the impending threat of climate change, carbon emissions is sometimes stated as a 'carbon dioxide equivalent (CO_2-e) or CO_2 concentrations in parts per million (ppm)'. For this story, it would be impractical to continually make a technical distinction between these terms or speculate on their intended meaning as used by the referenced authors. Hence, greenhouse gas emissions and carbon emissions (or CO_2) confer the same consequences with respect to climate change and global warming unless otherwise specified.

Climate change although framed and reframed in a myriad of ways, climate change is, and perhaps always was, a human thinking and acting problem. From a technical perspective, it is a change in the climate that occurs over a period of decades or more that can be identified and measured. Climate change is a consequence of several influences, including natural internal processes and external forcings to anthropogenic changes in the composition of the atmosphere, deforestation and patterns of land use. Moreover, climate change is often used interchangeably with global warming. The term global warming, however, is somewhat misleading because anthropogenic climate change might result in both a warming and cooling of the planet. Global warming refers to an increase in the Earth's mean surface temperature; whereas, climate change encompasses long-term climatic changes in the Earth's temperature, wind-patterns, precipitation, and sea levels, among other things.

Framing (emphasis framing) to consciously or unconsciously sway an audience's preferences by understating, overstating, misstating, and/or omitting, particular points of reference in a communication to create a preferred impression which, therefore, has ethical implications. Moreover, the strength of a framing effect is a function of the strength of the impressionability and/or the conviction of an audience's preferences.

Mental models personally formed, internally held cognitive representations that have, to a mutable extent, an analogical coherence with the external physical world – notwithstanding that gaps, or blind spots, might exist and values and beliefs might exert an influence – that affect a person's thinking and acting.

Planned retreat (also called a managed retreat or strategic retreat) is defined as a precautionary climate change adaptation policy that involves a sequence of systematically phased processes and actions designed to withdraw physical assets from the coastline and establish an ecological and/or structural buffer of a tractable dimension. The purpose of which is to protect against extreme climate events, such as storm and tidal surge, that have the potential to cause excessive erosion, substantial damage to property and infrastructure, and loss of life. Climate change is predicted to intensify extreme climate events. Knowing the precise time to withdraw – a socioeconomic 'duty of care' – is the definitive question.

Scientific phenomena in the traditional sense, observable and extraordinary phenomena that are discovered in nature and require an explanation. Alternatively, the discovery of the phenomena comes post the theory that the phenomena exist; for example, climate change.

Scientific reality for the purpose herein, is the empirical model of climate change held by the IPCC, notwithstanding the uncertainty inherent in the science.

Sea level rise the consequence of, among other things, rising ocean heat and thus ocean thermal expansion, the melting of land ice such as glaciers, ice caps, and the major ice sheets of Greenland and Antarctica, as well as groundwater extraction and reservoir impoundment.

Storm surge the momentary increase in sea height at a particular time and place due to extreme climatic conditions such as low atmospheric pressure and/or strong winds. The storm surge itself is defined as the excess above the expected tide at that particular locale, which is also a function of coastal topography and near-shore bathymetry.

Zombie for the purpose herein, the mindless enterprise of human behaviour in its unconscious endeavour to simulate what it means to be human.

References

Aalberg, T., Blekesaune, A., & Elvestad, E. (2013). Media choice and informed democracy: Toward increasing news consumption gaps in Europe? *The International Journal of Press/Politics, 18*(3), 281–303. doi: 10.1177/1940161213485990

Abbott, T. (2017). *Transcript of the Hon. Tony Abbott MP, address to the Global Warming Policy Foundation, Westminster, London.* October 10, 2017. Retrieved 28 January, 2018, from http://tonyabbott.com.au/2017 /10transcript-hon-tony-abbott-mp-address-global-warming-policy-foundation-westminster-london/.

Abelson, R. P. (1964). Mathematical models of the distribution of attitudes under controversy. In N. Frederiksen & H. Gulliksen (Eds.), *Contributions to mathematical psychology* (pp. 142–160). New York, NY: Holt, Rinehart, and Winston.

Abelson, R. P. (2014). Attitude extremity. In R. E. Petty & J. A. Krosnick (Eds.), *Attitude strength: Antecedents and consequences* (pp. 25–41). New York, NY: Psychology Press.

Ackoff, R. L. (1999). On learning and the systems that facilitate it. *Reflections, 1*(1), 14–24. doi: 10.1162/152417399570250

Adler, C. E., & Hadorn, G. H. (2014). The IPCC and treatment of uncertainties: Topics and sources of dissensus. *Wiley Interdisciplinary Reviews: Climate Change, 5*(5), 663–676. doi: 10.1002/wcc.297

Ahmed, S. H., Guillem, K., & Vandaele, Y. (2013). Sugar addiction: Pushing the drug-sugar analogy to the limit. *PubMed, 16*(4), 434–439. doi: 10.1097/MCO.0b013e328361c8b8

Aizawa, K. (2016). Embedded, embodied, extended, and enactive cognition: A Copernican revolution in psychology? In B. P. McLaughlin (Ed.), *Philosophy: Mind* (pp. 395–416). New York, NY: Macmillan.

Alexander, P. A. (2016). The arguments for and the reasoning about epistemic cognition. In J. A. Greene, W. A. Sandoval, & I. Bråten (Eds.), *Handbook of epistemic cognition* (pp. 100–111). New York, NY: Routledge.

Allahverdyan, A. E., & Galstyan, A. (2014). Opinion dynamics with confirmation bias. *PLoS ONE, 9*(7). doi: 10.1371/journal.pone.0099557

Alley, R. B., Clark, P. U., Huybrechts, P., & Joughin, I. (2005). Ice-sheet and sea-level changes. *Science, 310*(5747), 456–460. doi: 10.1126/science.1114613

Allison, I., Alley, R. B., Fricker, H. A., Thomas, R. H., & Warner, R. C. (2009). Ice sheet mass balance and sea level. *Antarctic Science, 21*(5), 413–426. doi: 10.1017/S0954102009990137

Altheide, D. L. (2017). *Terrorism and the politics of fear* (2nd edn). New York, NY: Rowman & Littlefield.

American Psychiatric Association. (2013). *Diagnostic and statistical manual of mental disorders: DSM-5* (5th edn). Arlington, VA: APA.

Anderson, J. E. (2014). *Public policymaking* (8th edn). Stamford, CT: Cengage Learning.

Ansell, C., & Boin, A. (2017). Taming deep uncertainty: The potential of pragmatist principles for understanding and improving strategic crisis management. *Administration & Society*, 1–34. doi: 10.1177/0095399717747655

Arcilla, R. V. (2002). Modernising media or modernist medium? The struggle for liberal learning in our information age. *Journal of Philosophy of Education, 36*(3), 457–465. doi: 10.1111/1467-9752.00287

Arendt, H. (1969). *On violence.* London, UK: Harcourt.

Arendt, H. (2005). Truth in politics. In D. Wood & J. Medina (Eds.), *Truth: Engagements across philosophical traditions* (pp. 295–314). Malden, MA: Blackwell.

Arnold, R. D. (1990). *The logic of congressional action.* New York, NY: Yale University Press.

Ashare, R. L., Norris, C. J., Wileyto, E. P., Cacioppo, J. T., & Strasser, A. A. (2013). Individual differences in positivity offset and negativity bias: Gender-specific associations with two serotonin receptor genes. *Personality and Individual Differences, 55*(5), 469–473. doi: 10.1016/j.paid.2013.04.009

Ashton, P., & Christophers, B. (2015). On arbitration, arbitrage and arbitrariness in financial markets and their governance: Unpacking LIBOR and the LIBOR scandal. *Economy and Society, 44*(2), 1–30. doi: 10.1080/03085147.2015.1013352

Attfield, R. (2014). *Environmental ethics*. Cambridge, UK: Polity Press.

Bach, J., & Gelber, D. (Writers) & J. Cameron, J. Weintraub, & A. Schwarzenegger (Directors). (2014). *Years of living dangerously*. New York, NY: Showtime Networks.

Bacon, F. (1605). *The advancement of learning*. Oxford, UK: The Clarendon Press.

Baddeley, A. (2000). The episodic buffer: A new component of working memory? *Trends in Cognitive Sciences, 4*(11), 417–423. doi: 10.1016/S1364-6613(00)01538-2

Baddeley, A., & Hitch, G. (1974). Working memory. *Psychology of Learning and Motivation: Advances in Research and Theory, 8*, 47–89.

Bailey, A., Giangola, L., & Boykoff, M. T. (2014). How grammatical choice shapes media representations of climate (un)certainty. *Environmental Communication: A Journal of Nature and Culture, 8*(2), 197–215. doi: 10.1080/17524032.2014.906481

Ball, S. J. (2012). Performativity, commodification and commitment: An I-Spy guide to the neoliberal university. *British Journal of Educational Studies, 60*(1), 17–28. doi: 10.1080/00071005.2011.650940

Barrett, R. (2012). No climate change in school: LNP. *The Australian*. Retrieved 17 July, 2013, from www.theaustralian.com.au/national-affairs/state-politics/no-climate-change-in-school-lnp/story-e6frgczx-1226425782372.

Bartlett, F. C. (1932). *Remembering: A study in experimental and social psychology.* London, UK: Cambridge University Press.

Bast, E., Doukas, A., Pickard, S., van der Burg, L., & Whitely, S. (2015). Empty promises: G20 subsidies to oil, gas and coal production. Washington, DC: Oil Change International Overseas Development Institution.

Baston, D. (2014). *The altruism question: Toward a social-psychological answer*. New York, NY: Psychology Press.

Bates, G. (2013). *Environmental law in Australia*. Chatswood, Australia: LexisNexis Butterworths.

Baudrillard, J. (1990). *Cool memories* (C. Turner, Trans.). New York, NY: Verso.

Baudrillard, J. (1994). *Simulacra and simulation* (S. Faria Glaser, Trans.). Ann Arbor, MI: University of Michigan.

Bauman, Z. (2001). *The individualized society*. Cambridge, UK: Polity Press.

Baumeister, R. F., Bratslavsky, E., Finkenauer, C., & Vohs, K. D. (2001). Bad is stronger than good. *Review of General Psychology, 5*(4), 323–370. doi: 10.1037/1089-2680.5.4.323

Bechtel, M. M., Hainmueller, J., Hangartner, D., & Hebling, M. (2015). Reality bites: The limits of framing effects for salient and contested policy issues. *Political Science Research and Methods, 3*(3), 683–695. doi: 10.1017/psrm.2014.39

Beck, S., Borie, M., Chilvers, J., Esguerra, A., Heubach, K., Hulme, M., … Görg, C. (2014). Towards a reflexive turn in the governance of global environmental expertise: The cases of the IPCC and the IPBES. *Gaia: Ökologische Perspektiven für Wissenschaft und Gesellschaft, 23*(2), 80–87.

Beck, U. (2010). Climate for change, or how to create a green modernity? *Theory, Culture and Society, 27*(2–3), 254–266. doi: 10.1177/0263276409358729

Beck, U. (2013). *World at risk* (C. Cronin, Trans.). Cambridge, UK: Polity Press.

Beck, U. (2014). How climate change might save the world. *Development and Society, 43*(2), 169–183.

Beck, U. (2015). Emancipatory catastrophism: What does it mean to climate change and risk society? *Current Sociology, 63*(1), 75–88. doi: 10.1177/0011392114559951

Beeson, M., & McDonald, M. (2013). The politics of climate change in Australia. *Australian Journal of Politics & History, 59*(3), 331–348. doi: 10.1111/ajph.12019

Bentham, J. ([1821] 2004). *Jeremy Bentham's economic writings: Critical edition based on his printed works and unprinted manuscripts* (Vol. 1). New York, NY: Routledge.

Bentley, T. (2014). Design in policy: Challenges and sources of hope for policymakers. In C. Bason (Ed.), *Design for policy* (pp. 11–22). Farnham, UK: Gower.

Berger, P. L., & Luckmann, T. (1967). *The social construction of reality: A treatise in the sociology of knowledge*. New York, NY: Penguin.

Bernacer, J., Balderas, G., Martinez-Valbuena, I., Pastor, M. A., & Murillo, J. I. (2014). The problem of consciousness in habitual decision making. *The Behavioral and Brain Sciences, 37*(1), 21–22. doi: 10.1017/s0140525x13000642

Bernstein, A. E. (2017). *Hell and its rivals*. London, UK: Cornell University Press.

Bertolotti, M., Catellani, P., Douglas, K. M., & Sutton, R. M. (2013). The 'big two' in political communication: The effects of attacking and defending politicians' leadership or morality. *Social Psychology, 44*(2), 117–128. doi: 10.1027/1864-9335/a000141

Bertsch, S., & Pesta, B. J. (2009). The Wonderlic Personnel Test and elementary cognitive tasks as predictors of religious sectarianism, scriptural acceptance and religious questioning. *Intelligence, 37*(3), 231–237. doi: 10.1016/j.intell.2008.10.003

Besley, J. C., & Nisbet, M. (2013). How scientists view the public, the media and the political process. *Public Understanding of Science, 22*(6), 644–659. doi: 10.1177/0963662511418743

Bessette, D. L., Mayer, L. A., Cwik, B., Vezér, M., Keller, K., Lempert, R. J., & Tuana, N. (2017). Building a values-informed mental model for New Orleans climate risk management. *Risk Analysis, 37*(10), 644–659. doi: 10.1111/risa.12743

Beven, K. (2012). Causal models as multiple working hypotheses about environmental processes. *Comptes Rendus Geoscience, 344*(2), 77–88. doi: 10.1016/j.crte.2012.01.005

Beven, K. (2013). So how much of your error is epistemic? Lessons from Japan and Italy. *Hydrological Processes, 27*(11), 1677–1680. doi: 10.1002/hyp.9648

Biermann, F. (2014). *Earth system governance: World politics in the Anthropocene*. Cambridge, MA: MIT Press.

Bhave, A. G., Conway, D., Dessai, S., & Stainforth, D. A. (2016). Barriers and opportunities for robust decision making approaches to support climate change adaptation in the developing world. *Climate Risk Management, 14*, 1–10. doi: 10.1016/j.crm.2016.09.004

Birkland, T. A. (2015). *An introduction to the policy process* (3rd edn). New York, NY: Routledge.

Björkman, M. (1984). Decision making, risk taking and psychological time: Review of empirical findings and psychological theory. *Scandinavian Journal of Psychology, 25*(1), 31–49.

Blainey, G. (2015). *The story of Australia's people: The rise and fall of ancient Australia*. Scoresby, Australia: Penguin.

Blühdorn, I. (2007). Sustaining the unsustainable: Symbolic politics and the politics of simulation. *Environmental Politics, 16*(2), 251–275. doi: 10.1080/09644010701211759

Boehmer-Christiansen, S. (1994). The precautionary principle in Germany – enabling government. In T. O'Riordan & J. Cameron (Eds.), *Interpreting the precautionary principle* (pp. 31–60). London, UK: Earthscan.

Bohm, D., & Nichol, L. (2013). *On dialogue*. Hoboken, NJ: Routledge.

Boholm, A., & Corvellec, H. (2014). A relational theory of risk: Lessons for risk communication. In J. Arvai & L. Rivers (Eds.), *Effective risk communication* (pp. 6–22). London, UK: Routledge.

Borden, M. E. (2017). *Psychology in the light of the East*. London, UK: Rowman & Littlefield.

Boschetti, F. (2012). A computational model of a mental model used to reason about climate change. *Environmental Science & Policy, 15*(1), 125–135. doi: 10.1016/j.envsci.2011.09.006

Boulding, E. K. ([1966] 2013). The economics of the coming spaceship Earth. In H. Jarrett (Ed.), *Environmental quality in a growing economy: Essays from the sixth RFF forum* (pp. 3–14). Hoboken, NJ: Taylor & Francis.

Bourke, L. (2017, October 10.). Tony Abbott says climate change 'could be beneficial' because people die in the cold. *Sydney Morning Herald.* Retrieved 25 January, 2018, from www.smh.com.au/federal-politics/political-news/tony-abbott-says-climate-change-could-be-beneficial-because-more-people-die-20171009-gyxi9y.html.

Boyd, R., Turner, J. C., & Ward, B. (2015). Tracking intended nationally determined contributions: What are the implications for greenhouse gas emissions in 2030? *ESRC Centre for Climate Change Economics and Policy (CCCEP) and Grantham Research Institute on Climate Change and the Environment*, 1–19.

Boykoff, M. T. (2011). *Who speaks for the climate? Making sense of media reporting on climate change.* New York, NY: Cambridge University Press.

Boylan, M. (2013). *Environmental ethics*. Malden, MA: Wiley Blackwell.

Bradshaw, J. L. (2014). *Human evolution*. Hoboken, NJ: Taylor & Francis.

Brooks, M. (2006). *World war z: An oral history of the zombie war*. New York, NY: Crown Publishing.

Brooks, M. (2011). *The zombie survival guide*. New York, NY: Random House.

Brulle, R. J. (2014). Institutionalizing delay: Foundation funding and the creation of U.S. climate change counter-movement organizations. *Climatic Change, 122*(4), 681–694. doi: 10.1007/s10584-013-1018-7

Bruner, J. S. (2009). *Actual minds, possible worlds*. Cambridge, MA: Harvard University Press.

Brunswick, E. (1952). *The conceptual framework of psychology*. Chicago, IL: University of Chicago Press.

Brunswick, E. (1956). *Perception and the representative design of psychological experiments*. Berkley, CA: University of California Press.

Bryant, N. (2014). *The rise and fall of Australia: How a great nation lost its way*. North Sydney, Australia: Random House.

Brynjolfsson, E., & McAfee, A. (2016). *The Second machine Age: Work, progress, and prosperity in a time of brilliant technologies*. New York, NY: W. W. Norton & Company.

Buchan, B., & Hill, L. (2014). *An intellectual history of political corruption*. Hampshire, UK: Palgrave Macmillan.

Buchanan, J. M., & Tullock, G. (1965). *The calculus of consent: Logical foundations of constitutional democracy* (Vol. 100). Ann Arbor, MI: University of Michigan Press.

Budescu, D. V., Broomell, S., & Por, H.-H. (2009). Improving communication of uncertainty in the reports of the Intergovernmental Panel on Climate Change. *Psychological Science, 20*(3), 299–308. doi: 10.1111/j.1467-9280.2009.02284.x

Budescu, D. V., Por, H.-H., & Broomell, S. (2012). Effective communication of uncertainty in the IPCC Reports. *Climatic Change, 113*(2), 181–200. doi: 10.1007/s10584-011-0330-3

Bull, H., & Watson, A. (Eds.) (1984) *The expansion of international society*. Oxford, UK: Clarendon Press.

Buss, D. M. (2008). *Evolutionary psychology: The new science of the mind* (3rd edn). Boston, MA: Allyn and Bacon.

Butt, P. J. (2004). *Butterworths concise Australian legal dictionary* (3rd edn). Chatswood, Australia: LexisNexis.

Buttel, F. H. (2003). Environmental sociology and the explanation of environmental reform. *Organization & Environment, 16*(3), 306–344. doi: 10.1177/1086026603256279

Cacciatore, M. A., Scheufele, D. A., & Iyengar, S. (2015). The end of framing as we know it … and the future of media effects. *Mass Communication and Society, 1*(1), 1–17. doi: 10.1080/15205436.2015.1068811

Cacioppo, J. T., Cacioppo, S., & Petty, R. E. (2017). The neuroscience of persuasion: A review with an emphasis on issues and opportunities. *Social Neuroscience, 13*(2), 129–172. doi: 10.1080/17470919.2016.1273851

Cacioppo, J. T., & Freberg, L. (2014). *Discovering psychology: The science of mind*. Boston, MA: Cengage Learning.

Calaprice, A. (2000). *The expanded quotable Einstein*. Princeton, NJ: Princeton University Press.

Calaprice, A. (Ed.) (2011). *The ultimate quotable Einstein*. Princeton, NJ: Princeton University Press.

Callendar, G. S. (1938). The artificial production of carbon dioxide and its influence on temperature. *Quarterly Journal of the Royal Meteorological Society, 64*(275), 223–240.

Cameron, J. (Writer). (2009). *Avatar*. Los Angeles, CA: 20th Century Fox.

Capra, F., & Luisi, P. L. (2014). *The systems view of life: A unifying vision*. New York, NY: Cambridge University Press.

Carah, N., & Louw, E. (2015). *Media and society: Production, content and participation*. Los Angeles, CA: Sage.

Carli, L. L. (2017). Social influence and gender. In S. G. Harkins, K. D. Williams, & J. M. Burger (Eds.), *The Oxford handbook of social influence* (pp. 33–51). New York, NY: Oxford University Press.

Carlson, M. (2017). *Journalistic authority*. New York, NY: Columbia University Press.

Carlson, W. B. (2005). Inventor of dreams. *Scientific American, 292*, 78–85. doi: 10.1038/scientificamerican0305-78

Carr, N. (2008). Is Google making us stupid? *Yearbook of the National Society for the Study of Education, 107*(2), 89–94. doi: 10.1111/j.1744-7984.2008.00172.x

CCP Research Foundation. (2018). *Conduct costs project report 2017*. England & Wales, UK. CCP Research Foundation. Retrieved 8 February, 2018, from http://foreigners.textovirtual.com/ccp-research-foundation/271/221503/conduct-costs-project-report-pr-no-1-aug-2017.pdf.

Chalmers, D. J. (1996). *The conscious mind: In search of a fundamental theory*. New York, NY: Oxford University Press.

Chalmers, D. J. (2010). *The character of consciousness*. Oxford, UK: Oxford University Press.

Chang, S. E., Yip, J. Z. K., Conger, T., Oulahen, G., & Marteleira, M. (2018). Community vulnerability to coastal hazards: Developing a typology for disaster risk reduction. *Applied Geography, 91*, 81–88. doi: 10.1016/j.apgeog.2017.12.017

Chen, X. (2011). Why do people misunderstand climate change? Heuristics, mental models and ontological assumptions. *Climatic Change, 108*(1–2), 31–46. doi: 10.1007/s10584-010-0013-5

Chiu, K., Grundy, Q., & Bero, L. (2017). 'Spin' in published biomedical literature: A methodological sytematic review. *PLOS Biology,* 15(19), e2002173. doi: 10.1371/journal.pbio.2002173

Chong, D., & Druckman, J. N. (2007a). Framing public opinion in competitive democracies. *American Political Science Review, 101*(4), 637–655. doi: 10.1017/s0003055407070554

Chong, D., & Druckman, J. N. (2007b). Framing theory. *Annual Review of Political Science, 10*, 103–126. doi: 10.1146/annurev.polisci.10.072805.103054

Chong, D., & Druckman, J. N. (2010). Dynamic public opinion: Communication effects over time. *American Political Science Review, 104*(4), 663–680. doi: 10.1017/S0003055410000493

Ciplet, D., Roberts, J. T., & Khan, M. R. (2015). *Power in a warming world: The new global politics of climate change and the remaking of environmental inequality*. Cambridge, MA: MIT Press.

Clark, A. (2013). Whatever next? Predictive brains, situated agents, and the future of cognitive science. *The Behavioral and Brain Sciences, 36*(3), 181–253. doi: 10.1017/S0140525X12000477

Clark, P. U., Dyke, A. S., Shakun, J. D., Carlson, A. E., Clark, J., Wohlfarth, B., … Marshall, A. (2009). The last glacial maximum. *Science, 325*(5941), 710–714. doi: 10.1126/science.112873

Clarke, A. C. ([1962] 2013). *Profiles of the future*. London, UK: Orion Publishing.

CleanGovBiz. (2014). *The rationale for fighting corruption: Background brief.* Paris, France: Organisation for Economic Co-operation and Development (OECD) Retrieved 14 February, 2016, from www.oecd.org/cleangovbiz/49 693613.pdf.

Cocking, J., & Murray, P. (2013). *Imagination*. London, UK: Taylor & Francis.

Colaiaco, J. A. (2013). *Socrates against Athens: Philosophy on trial*. Hoboken, NJ: Taylor & Francis.

Comer, R. J. (2013). *Abnormal psychology* (8th edn). New York, NY: Worth Publishers.

Conners, N., Petersen, L. C., & DiCaprio, L. (Writers) & N. Conners & L. C. Petersen (Directors). (2007). *The 11th hour*. Burbank, CA: Warner Independent Pictures.

Conway, R. (1971). *The great Australian stupor: An interpretation of the Australian way of life*. Melbourne, Australia: Sun Books.

Costello, F., & Watts, P. (2014). Surprisingly rational: Probability theory plus noise explains biases in judgment. *Psychological Review, 121*(3), 463–480. doi: 10.1037/a0037010

Cox, F., Huntingford, C., & Williamson, M. S. (2018). Emergent constraint on equilibrium climate sensitivity from global temperature variability. *Nature, 553*, 319–322. doi: 10.1038/nature25450

Craik, K. J. W. (1943). *The nature of explanation*. Cambridge, UK: Cambridge University Press.

Crouch, C. (2004). *Post-democracy*. Cambridge, UK: Polity Press.

Crupi, V., & Elia, F. (2017). Understanding and improving decisions in clinical medicine (I): Reasoning, heuristics, and error. *Internal and Emergency Medicine, 12*(5). doi: 10.1007/s1173

CSIRO. (2012). *Productivity Commission draft report: Barriers to effective climate change adaptation*. Clayton, Australia. CSIRO Publishing. Retrieved 19 October, 2017, from www.pc.gov.audata/assets/pdf_file/0016/117511 /subdr136.pdf.

Curran, D. (2009). Wicked. *Alternatives Journal, 35*(5), 8–11.

Dangendorf, S., Marcos, M., Wöppelmann, G., Conrad, C. P., Frederiske, T., & Riva, R. (2017). Reassessment of 20th century global mean sea level rise. *Proceedings of the National Academy of Sciences of the United States of America,* 14(23), 1–6. doi: 10.1073/pnas.1616007114

Dai, D. Y. (2017). A history of giftedness: A century of quest for identity. In S. I. Pfeiffer (Ed.), *APA handbook of giftedness and talent* (pp. 3–23). Washington, DC: American Psychological Association.

Dardot, P., & Laval, C. (2013). *The new way of the world: On neoliberal society*. Brooklyn, NY: Verso.

Dark, S. M., & Burgin, S. (2017). An examination of the efficacy of the precautionary principle as a robust environmental planning and management protocol. *Journal of Environmental Planning and Management, 60*(12), 2122–2132. doi: 10.1080/09640568.2016.1276436

Darwin, C. R. ([1839] 2008). *The voyage of the Beagle*. Auckland, New Zealand: Floating Press.

Davidson, A. (Writer). (2010). *The walking dead*. New York, NY: American Movie Classics (AMC).

Davies, M. (2011). Introduction to the special issue on critical thinking in higher education. *Higher Education Research and Development, 30*(3), 255–260. doi: 10.1080/07294360.2011.562145

Davis, J. L., & Vinogradova, N. T. (2017). Causes of accelerating sea level on the East Coast of North America. *Geophysical Research Letters, 44*(10), 5133–5141. doi: 10.1002/2017GL072845

Dawkins, R. (2006a). *The God delusion*. Boston, MA: Houghton Mifflin.

Dawkins, R. (2006b). *The selfish gene*. Oxford, UK: Oxford University Press.

Dawson, M. R. W. (2013). *Mind, body, world: Foundations of cognitive science*. Edmonton, Canada: Athabasca University Press.

de Botton, A. (2014). *The news*. London, UK: Penguin.

Denzau, A. T., & North, D. C. (1994). Shared mental models: Ideologies and institutions. *Kyklos, 47*(1), 3–31. doi: 10.org/10.1111/j.1467-6435.1994.tb02246.x

Der Kiureghian, A., & Ditlevsen, O. (2009). Aleatory or epistemic? Does it matter? *Structural Safety, 31*(2), 105–112. doi: 10.1016/j.strusafe.2008.06.020

DeSalle, R., & Tattersall, I. (2012). *The brain: Big bangs, behaviors, and beliefs*. New Haven, CT: Yale University Press.

Descartes, R. ([1641] 2008). *Meditations on first philosophy: With selections from the objections and replies*. (M. Moriarty, Trans.). Oxford, UK: Oxford University Press.

Deser, C., Phillips, A., Bourdette, V., & Teng, H. (2012). Uncertainty in climate change projections: The role of internal variability. *Climate Dynamics, 38*(3), 527–546. doi: 10.1007/s00382-010-0977-x

Deuze, M. (2013). Living as a zombie in media (is the only way to survive). *MATRIZes, 7*(2), 113–129. doi: 10.11606/issn.1982-8160.v7i2p113-129

Deville, A., & Harding, R. (1997). *Applying the precautionary principle*. Annandale, Australia: Federation Press.

Dewey, J. (1922). Public opinion (review). *The New Republic, 30*, 286–288.

Dewey, J. (1927). The role of philosophy in the history of civilization. *The Philosophical Review, 36*(1), 1–9.

Dewey, J. ([1927] 2012). *The public and its problems*. University Park, PA: Pennsylvania State University Press.

Dewey, J. (1933). *How we think*. New York, NY: Dover Publications.

Dexter, E. G. (1899). Conduct and the weather: An inductive study of the mental effects of definite meteorological conditions. *Psychological Review, 6*(5), 539–540. doi: 10.1037/h0065689

Dieng, H. B., Cazenave, A., & Ablain, M. (2017). New estimate of the current rate of sea level rise from sea level budget approach. *Geophysical Research Letters, 44*(8), 539–540. doi: 10.1002/2017GL073308

Dietrich, E. (2016). The philosophy of artificial intelligence. In B. P. McLaughlin (Ed.), *Philosophy: Mind* (pp. 375–394). New York, NY: Macmillan.

Dilthey, W. (1988). *Introduction to the human sciences: An attempt to lay a foundation for the study of society and history* (J. R. Betanzos, Trans.). Detroit, MI: Wayne State University.

Dittrich, R., Wreford, A., & Moran, D. (2016). A survey of decision-making approaches for climate change adaptation: Are robust methods the way forward? *Ecological Economics, 122*, 79–89. doi: 10.1016/j.ecolecon.2015.12.006

Doan, M. D. (2014). Climate change and complacency. *Hypatia, 29*(3), 634–650. doi: 10.1111/hypa.12063

Dominik, A. (Writer). (2012). *Killing them softly*. Los Angeles, CA: Annapurna Pictures.

Donsbach, W., & Traugott, M. W. (2008). *The Sage handbook of public opinion research*. London, UK: Sage.

Doris, J. M. (2015). *Talking to our selves: Reflection, ignorance, and agency* (1st edn). New York, NY: Oxford University Press.

Drews, S., Antal, M., & van Den Bergh, J. C. J. M. (2018). Challenges in assessing public opinion on economic growth versus environment: Considering European and US data. *Ecological Economics, 146*, 265–272. doi: 10.1016/j.ecolecon.2017.11.006

Druckman, J. N. (2001a). The implications of framing effects for citizen competence. *Political Behavior, 23*(3), 225–256. doi: 10.1023/A:1015006907312

Druckman, J. N. (2001b). On the limits of framing effects: Who can frame? *The Journal of Politics, 63*(4), 1041–1066. doi: 10.1111/0022-3816.00100

Druckman, J. N. (2001c). Using credible advice to overcome framing effects. *Journal of Law, Economics and Organization, 17*(1), 62–82. doi: 10.1093/jleo/17.1.62

Druckman, J. N. (2004). Political preference formation: Competition, deliberation, and the (ir)relevance of framing effects. *The American Political Science Review, 98*(4), 671–686. doi: 10.1017/S0003055404041413

Druckman, J. N., & Lupia, A. (2017). Using frames to make scientific communication more effective. In K. H. Jamieson, D. Kahan, & D. A. Scheufele (Eds.), *The Oxford handbook of the science of science communication* (pp. 351–360). New York, NY: Oxford University Press.

Druckman, J. N., & Nelson, K. R. (2003). Framing and deliberation: How citizens' conversations limit elite influence. *American Journal of Political Science, 47*(4), 729–745. doi: 10.2307/3186130

Dub, R. (2016). Emotions. In B. P. McLaughlin (Ed.), *Philosophy: Mind* (pp. 245–266). New York, NY: Macmillan.

Dunlap, R. E. (2013). Climate change skepticism and denial: An introduction. *American Behavioral Scientist, 57*(6), 691–698. doi: 10.1177/0002764213477097

Dunlap, R. E., & Brulle, R. J. (2015). *Climate change and society: Sociological perspectives*. New York, NY: Oxford University Press.

Dupré, J. (2012). *Processes of life: Essays in the philosophy of biology*. New York, NY: Oxford University Press.

Duthel, H. (2011). *Rupert Murdoch: The politico media complex mogul*. New York, NY: CreateSpace.

Dutton, A., Carlson, A. E., Long, A. J., Milne, G. A., Clark, P. U., DeConto, R., … Raymo, M. E. (2015). Sea-level rise due to polar ice-sheet mass loss during past warm periods. *Science, 349*(6244). doi: 10.1126/science.aaa4019

Dutton, A., & Lambeck, K. (2012). Ice volume and sea level during the last interglacial. *Science, 337*(6091), 216–219. doi: 10.1126/science.1205749

Earnest, C. E. (2015). The gardener and the craftsman: Four types of complexity in global life. In E. Kavalski (Ed.), *World politics at the edge of chaos: Reflections on complexity and global life* (pp. 31–51). Albany, NY: State University of New York.

Edgerly, S. (2015). Red media, blue media, and purple media: News repertoires in the colorful media landscape. *Journal of Broadcasting and Electronic Media, 59*(1), 1–21. doi: 10.1080/08838151.2014.998220

Edwards, L. (2013). *The passion of politics: The role of ideology and political theory in Australia*. Sydney, Australia: Allen & Unwin.

Elshof, L. (2010). Changing worldviews to cope with a changing climate. In R. Irwin (Ed.), *Climate change and philosophy: Transformational possibilities* (pp. 75–108). New York, NY: Continuum.

Entman, R. M. (1993). Framing: Toward clarification of a fractured paradigm. *Journal of Communication, 43*(4), 51–58. doi: 10.1111/j.1460-2466.1993.tb01304.x

Esser, F., & Strömbäck, J. (2014). *Mediatization of politics: Understanding the transformation of Western democracies*. New York, NY: Palgrave Macmillan.

Evans, J. (1998). Matching bias in conditional reasoning: Do we understand it after 25 years? *Thinking and Reasoning, 4*(1), 45–110. doi: 10.1080/135467898394247

Evans, J. (2003). In two minds: Dual-process accounts of reasoning. *Trends in Cognitive Sciences, 7*(10), 454–459. doi: 10.1016/j.tics.2003.08.012

Evans, J. (2007). *Hypothetical thinking: Dual processes in reasoning and judgement* (Vol. 3). New York, NY: Psychology Press.

Evans, J. (2008). Dual-processing accounts of reasoning, judgment, and social cognition. *The Annual Review of Psychology, 59*, 255–278. doi: 10.1146/annurev.psych.59.103006.093629

Evans, J. (2010). *Thinking twice: Two minds in one brain.* Oxford, UK: Oxford University Press.

Evans, J. (2014a). Rationality and the illusion of choice. *Frontiers in Psychology, 5*(104), 1–4. doi: 10.3389/fpsyg.2014.00104

Evans, J. (2014b). Two minds rationality. *Thinking and Reasoning, 20*(2), 129–146. doi: 10.1080/13546783.2013.845605

Evans, J., & Frankish, K. (2009). *In two minds: Dual processes and beyond.* New York, NY: Oxford University Press.

Farmer, G. T., & Cook, J. (2013). Earth's energy budget. In G. T. Farmer & J. Cook (Eds.), *Climate change science: A modern synthesis: Volume 1 – the physical climate* (pp. 81–95). New York, NY: Springer.

Farrell, J. (2015). Network structure and influence of the climate change counter-movement. *Nature Climate Change, 6*(4), 370–374. doi: 10.1038/nclimate2875

Farrell, J., & Rabin, M. (1996). Cheap talk. *The Journal of Economic Perspectives, 10*(3), 103–118.

Fay, A. J., & Maner, J. K. (2014). When does heat promote hostility? Person by situation interactions shape the psychological effects of haptic sensations. *Journal of Experimental Social Psychology, 50*, 210–216. doi: 10.1016/j.jesp.2013.10.006

Fazio, R. H. (1990). Multiple processes by which attitudes guide behavior: The MODE model as an integrative framework. *Advances in Experimental Social Psychology, 23*, 75–109. doi: 10.1016/S0065-2601(08)60318-4

Fazio, R. H., & Towels-Schwen, T. (1999). The MODE model of attitude-behavior processes. In S. Chaiken & Y. Trope (Eds.), *Dual-process theories in social psychology* (pp. 97–116). New York, NY: Guilford Press.

Ferguson, N. (2013). *The great degeneration: How institutions decay and economies die.* London, UK: Penguin.

Festinger, L. (1962). *A theory of cognitive dissonance*. Stanford, CA: Stanford University Press.

Feyerabend, P. K. (1981). *Realism, rationalism, and scientific method* (Vol. 1.). New York, NY: Cambridge University Press.

Feynman, R. P. (1992). *'Surely you're joking Mr Feynman'!: Adventures of a curious character as told to Ralph Leighton.* London, UK: Vintage Books.

Figueres, C., Schellnhuber, H. J., Whiteman, G., Rockström, J., Hobley, A., & Rahmstorf, S. (2017). Three years to safeguard our climate. *Nature, 546*, 593–594.

Fiske, S. T. (2014). *Social beings: Core motives in social psychology*. Hoboken, NJ: John Wiley and Sons.

Fiske, S. T., & Durante, F. (2014). Never trust a politician? Collective distrust, relational accountability, and voter response. In J. W. van Prooijen & P. A. M. van Lange (Eds.), *Power, politics, and paranoia: Why people are suspicious of their leaders* (pp. 91–105). Cambridge, UK: Cambridge University Press.

Fiske, S. T., & Taylor, S. E. (2013). *Social cognition: From brains to culture* (2nd edn). London, UK: Sage.

Flanagan, D. P., & Dixon, S. G. (2014). The Cattell-Horn-Carroll Theory of cognitive abilities. *The Encyclopedia of Special Education* (article). doi: 10.1002/9781118660584.ese0431

Fleishman-Mayer, L. A., & de Bruin, W. B. (2014). The 'mental models' methodology for developing communications. In J. Arvai & L. Rivers (Eds.), *Effective risk communication* (pp. 165–189). London, UK: Routledge.

Flurry Analytics. (2017). *U.S. consumers time-spent on mobile crosses 5 hours a day.* New York, NY: Flurry Analytics. Retrieved March 6, 2017, from http://flurrymobile.tumblr.com/post/157921590345/us-consumers-time-spent-on-mobile-crosses-5.

Fodor, J. (2001). *The mind doesn't work that way: The scope and limits of computational psychology*. London, UK: MIT Press.

Foote, E. (1856). Circumstances affecting the heat of the sun's rays. In B. Silliman, B. Silliman Jr., and J. D. Dana (Eds.), *The American Journal of Science and Arts* (pp. 382–383) Washington, DC: The American Journal of Science and Arts.

Forgas, J. P., Vincze, O., & Laszlo, J. (2014). *Social cognition and communication.* London, UK: Psychology Press.

Forrester, J. W. (1971). Counterintuitive behavior of social systems. *Technological Forecasting & Social Change, 2*(2), 109–140. doi: 10.1016/S0040-1625(71)80001-X

Foucault, M. ([1966] 1970). *The order of things: An archaeology of the human sciences.* New York, NY: Pantheon.

Foucault, M. ([1989] 2012). *The order of things*. Hoboken, NJ: Taylor & Francis.

Fox, B. (2017). It's nature and nurture: Integrating biology and genetics into the social learning theory of criminal behavior. *Journal of Criminal Justice, 49*, 122–131. doi: 10.1016/j.jcrimjus.2017.01.003

Frankel, M. (1999). *The times of my life and my life with the times*. New York, NY: Random House.

Frankfurt, H. G. (2005). *On bullshit*. Princeton, NJ: Princeton University Press.

Fredal, J. (2011). Rhetoric and bullshit. *College English, 73*(3), 243–259.

Fredericks, S. E. (2014). *Measuring and evaluating sustainability: Ethics in sustainability indexes*. London, UK: Routledge.

Fredriksson, P. G., & Neumayer, E. (2014). Corruption and climate change policies: Do the bad old days matter? *Environmental and Resource Economics*, 1–19. doi: 10.1007/s10640-014-9869-6

French, S. (2016). *Philosophy of science: Key concepts* (2nd edn). London, UK: Bloomsbury.

Freud, S. ([1926] 2013). *On Freud's inhibitions, symptoms and anxiety*. London, UK: Karnac.

Freud, S. ([1960] 2001). The question of a Weltanschauung. In J. Strachey (Ed.), *The standard edition of the complete psychological works of Sigmund Freud* (pp. 158–182). London, UK: Vintage.

Freund, K., Kizimchuk, S., Zapasnik, J., Esteves, K., & Mewburn, I. (2018). A labour of love: A critical examination of the 'labour icebergs' of massive open online courses. In D. Lupton, I. Mewburn, & P. Thomson (Eds.), *The digital academic: Critical perspectives on digital technologies in higher education* (pp. 122–139). New York, NY: Routledge.

Friedmann, J. (2011). *Insurgencies: Essays in planning theory*. New York, NY: Routledge.

Funtowicz, S. O., & Ravetz, J. R. (1993). Science for the Post-Normal Age. *FUTURES, 25*(7), 739–755. doi: 10.1016/0016-3287(93)90022-L

Furedi, F. (2013). *Authority: A sociological history.* New York, NY: Cambridge University Press.

Gamson, W. A., & Modigliani, A. (1989). Media discourse and public opinion on nuclear power: A constructionist approach. *American Journal of Sociology, 95*(1), 1–37. doi: 10.1086/229213

Garnaut, R. (2008a). *The Garnaut climate change review: Final report*. Port Melbourne, Australia: Cambridge University Press.

Garnaut, R. (2008b). Will climate change bring an end to the platinum age? *Asian Pacific Economic Literature, 22*(1), 1–14. doi: 10.1111/j.1467-8411.2008.00208.x

Garnaut, R. (2011). *The Garnaut review 2011: Australia in the global response to climate change*. Cambridge, UK: Cambridge University Press.

Garner, B. A., & Black, H. C. (Eds.). (2014) *Black's law dictionary* (10th edn). St. Paul, MN: West.

Gauffroy, C., & Barrouillet, P. (2014). A developmental mental model theory of conditional reasoning. In H. Markovits (Ed.), *The developmental psychology of reasoning and decision-making* (pp. 63–83). East Sussex, UK: Psychology Press.

Gellatly, A. R. H. (1989). Human inference. In K. J. Gilhooly (Ed.), *Human and machine problem solving* (pp. 233–264). New York, NY: Plenum Press.

Giddens, A. (1994). Living in a post-traditional society. In U. Beck, A. Giddens, & S. Lash (Eds.), *Reflexive modernization: Politics, tradition and aesthetics in the modern social order* (pp. 56–109). Stanford, CA: Stanford University Press.

Giddens, A. ([1991] 2013). *The consequences of modernity*. Hoboken, NJ: Wiley Blackwell.

Gifford, R. (2011). The dragons of inaction: Psychological barriers that limit climate change mitigation and adaptation. *American Psychologist, 66*(4), 290–302. doi: 10.1037/a0023566

Gifford, R. (2014). Environmental psychology matters. *Psychology, 65*(1), 541–579. doi: 10.1146/annurev-psych-010213-115048

Glacken, C. J. (1967). *Traces on the Rhodian Shore: Nature and culture in Western thought from ancient times to the end of the eighteenth century*. Oakland, CA: University of California Press.

Global Carbon Project. (2017). *Global carbon budget: Summary highlights*. Canberra, Australia. Retrieved 29 January, 2018, from www.globalcarbonproject.org/carbonbudget/17/highlights.htm.

Glover, J. (2014). *Alien landscapes? Interpreting disordered minds*. Cambridge, UK: Harvard University Press.

Glynn, C. J., Herbst, S., Lindeman, M., O'Keefe, G. J., & Shapiro, R. Y. (2016). *Public opinion* (3rd edn). Boulder, CO: Westview Press

Goldacre, B. (2009). *Bad science*. London, UK: Harper Perennial.

Golding, W. (1954). *Lord of the flies*. London, UK: Faber & Faber.

Goodwin, B. (2001). *The philosophy of utopia*. Essex, UK: Frank Cass Publishers.

Goodwin, B. (2014). *Using political ideas* (6th edn). West Sussex, UK: Wiley.

Gore, A., & Guggenheim, D. (Writers) & D. Guggenheim (Director). (2006). *An inconvenient truth*. Los Angeles, CA: Paramount Pictures.

Gottfredson, L. S. (1997). Mainstream science on intelligence: An editorial with 52 signatories, history, and bibliography. *Intelligence, 24*(1), 13–23. doi: 10.1016/S0160-2896(97)90011-8

Gould, S. J. (1996). *The mismeasure of man* (Revised and expanded ed.). New York, NY: Norton.

Grace, K., Salvatier, J., Dafoe, A., Zhang, B., & Evans, O. (2017). When will AI exceed human performance? Evidence from AI experts. *Cornell University Library*, arXiv:1705.08807 [cs.AI]

Greenfield, S. (2015). *Mind change: How digital technologies are leaving their mark on our brains*. London, UK: Random House.

Griffin, E. A. (2009). *A first look at communication theory* (7th edn). New York, NY: McGraw-Hill.

Griffin, J. M., & Steele, H. B. (2013). *Energy economics and policy*. Orlando, FL: Academic Press.

Guivarch, C., Lempert, R., & Trutnevyte, E. (2017). Scenario techniques for energy and environmental research: An overview of recent developments to broaden the capacity to deal with complexity and uncertainty. *Environmental Modelling and Software, 97*, 201–210. doi: 10.1016/j.envsoft.2017.07.017

Gurevitch, Z. D. (1989). The power of not understanding: The meeting of conflicting identities. *The Journal of Applied Behavioral Science, 25*(2), 161–173. doi: 10.1177/0021886389252006

Guttmann, A. (1987). *Democratic education.* Princeton, NJ: Princeton University Press.

Guy, S., Kashima, Y., Walker, I., & O'Neill, S. (2014). Investigating the effects of knowledge and ideology on climate change beliefs. *European Journal of Social Psychology, 44*(5), 421–429. doi: 10.1002/ejsp.2039

Gwyther, D., Cook, S., Galton-Fenzi, B., Fraser, A., Graham, F., Kusahara, K., … van Omen, T. (2017). Position analysis: The Antartic ice sheet and sea level. *Antarctic Climate and Ecosystems Cooperative Research Centre*, 1–36. http://ecite.utas.edu.au/122498.

Habermas, J. ([1962] 1991). *The structural transformation of the public sphere: An inquiry into a category of bourgeois society* (T. Burger & F. Lawrence, Trans.). Cambridge, MA: MIT Press.

Haigh, Y. (2012). *Public policy in Australia: Theory and practice.* South Melbourne, Australia: Oxford University Press.

Hall, E. T. ([1959] 1980). *The silent language*. Westport, CT: Greenwood.

Hall, K., Curtin, A., & Rutherford, V. (2014). *Networks of mind: Brain, culture, learning.* New York, NY: Routledge.

Hall, M. (2015). *Exploring green crime: Introducing the legal, social and criminological contexts of environmental harm*. London, UK: Palgrave Macmillan.

Hall, M., & Farrall, S. (2013). The criminogenic consequences of climate change: Blurring the boundaries between offenders and victims. In A. Brisman & N. South (Eds.), *Routledge international handbook of green criminology* (pp. 26–58). New York, NY: Routledge.

Hall, S. (1997). Introduction. In S. Hall (Ed.), *Representation: Cultural representations and signifying practices* (pp. 1–12). London, UK: Sage.

Hamilton, C. (2011). Overheated debate. *New Scientist, 211*(2824), 28–29.

Hanna, E., Tulaczyk, S., Whitehouse, P. L., Zwally, H. J., Navarro, F. J., Pattyn, F., … Smith, B. (2013). Ice-sheet mass balance and climate change. *NATURE, 498*(7452), 51–59. doi: 10.1038/nature12238

Hansen, J., Sato, M., Hearty, P., Ruedy, R., Kelley, M., Masson-Delmotte, V., … Rignot, E. (2016). Ice melt, sea level rise and superstorms: Evidence from paleoclimate data, climate modeling, and modern observations that 2°C global warming is highly dangerous. *Atmospheric Chemistry and Physics Discussions, 15*(14), 20059–20179. doi: 10.5194/acpd-15-20059-2015

Hansson, S. O. (2018). Dealing with climate science denialism: Experiences from confrontations with other forms of pseudoscience. *Climate Policy*, 1752–7457. doi: 10.1080/14693062.2017.1415197

Harari, Y. N. (2011). *Sapiens: A brief history of humankind.* London, UK: Vintage Books.

Hardy, R. D., & Hauer, M. E. (2018). Social vulnerability projections improve sea-level rise risk assessments. *Applied Geography, 91*, 10–20. doi: 10.1016/j.apgeog.2017.12.019

Hardin, G. (1968). The tragedy of the commons. *Science, 162*(3859), 1243–1248. doi: 10.1126/science.162.3859.1243

Harris, P. G. (2015). What's wrong with climate politics. In S. Nicholson & P. K. Wapner (Eds.), *Global environmental politics: From person to planet* (pp. 137–143). Boulder, CO: Paradigm.

Hawking, S. W. (1988). *A brief history of time: From the big bang to black holes.* London, UK: Bantam Press.

Hawking, S. W., & Mlodinow, L. (2011). *The grand design.* London, UK: Transworld.

Haworth, A. (2012). *Understanding the political philosophers: From ancient to modern times.* London, UK: Routledge.

Hayek, F. A. (1960). *The constitution of liberty.* Chicago, IL: University of Chicago Press.

Head, B. W. (2014). Evidence, uncertainty, and wicked problems in climate change decision making in Australia. *Environment and Planning C: Government and Policy, 32*(4), 663–679. doi: 10.1068/c1240

Heal, G., & Millner, A. (2014). Uncertainty and decision making in climate change economics. *Review of Environmental Economics and Policy, 8*(1), 120–137. doi: 10.1093/reep/ret023

Heatley, B. (2017). Paris: Optimism, pessimism and realism. *Global Discourse, 7*(1), 10–22. doi: 10.1080/23269995.2017.1300402

Heisenberg, W. (1958). *Physics and philosophy.* New York, NY: Harper and Row.

Helm, D. (2010). Government failure, rent-seeking, and capture: The design of climate change policy. *Oxford Review of Economic Policy, 26*(2), 182–196. doi: 10.1093/oxrep/grq006

Herman, E. S., & Chomsky, N. (1994). *Manufacturing consent: The political economy of the mass media.* London, UK: Vintage.

Hertwig, R., & Gigerenzer, G. (1999). The 'conjunction fallacy' revisited: How intelligent inferences look like reasoning errors. *Journal of Behavioral Decision Making, 12*(4), 275–305. doi: 10.1002/(sici)1099–0771(199912)12:4<275::aid-bdm323>3.0.co;2-m

Heywood, A. (2014). *Global politics.* New York, NY: Palgrave Macmillan.

Hibbing, J. R., Smith, K. B., & Alford, J. R. (2014). Differences in negativity bias underlie variations in political ideology. *The Behavioral and Brain Sciences, 37*(3), 297–297. doi: 10.1017/s0140525x13001192

Hibbing, J. R., Smith, K. B., Peterson, J. C., & Feher, B. (2014). The deeper sources of political conflict: Evidence from the psychological, cognitive, and neuro-sciences. *Trends in Cognitive Sciences, 18*(3), 111–113. doi: 10.1016/j.tics.2013.12.010

Higgins, K. L. (2014). *Economic growth and sustainability: Systems thinking for a complex world.* Burlington, IL: Academic Press.

Higgins, P., Short, D., & South, N. (2013). Protecting the planet: A proposal for a law of ecocide. *Crime, Law and Social Change, 59*(3), 251–266. doi: 10.1007/s10611-013-9413-6

Hilton, I., & Kerr, O. (2017). The Paris Agreement: China's 'new normal' role in international climate negotiations. *Climate Policy, 17*(1), 45–58. doi: 10.1080/14693062.2016.1228521

Hobbes, T. ([1651] 1981). *Leviathan.* London, UK: Penguin.

Hodges, B. H. (2017). Conformity and divergence in interactions, groups, and culture. In S. G. Harkins, K. D. Williams, & J. M. Burger (Eds.), *The Oxford handbook of social influence* (pp. 87–106). New York, NY: Oxford University Press.

Holling, C. S., Gunderson, L., & Ludwig, D. (2002). In quest of a theory of adaptive change. In L. Gunderson (Ed.), *Panarchy: Understanding transformations in human and natural systems* (pp. 3–24). Washington, DC: Island Press.

Holmes, R. (2008). *The age of wonder: How the romantic generation discovered the beauty and terror of science*. London, UK: Harper Press.

Hostler, T. J., Wood, C., & Armitage, C. J. (2018). The influence of emotional cues on prospective memory: A systematic review with meta-analyses. *Cognition and Emotion*, 1–19. doi: 10.1080/02699931.2017.1423280

Houdé, O., Zago, L., Mellet, E., Moutier, S., Pineau, A., Mazoyer, B., & Tzourio-Mazoyer, N. (2000). Shifting from the perceptual brain to the logical brain: The neural impact of cognitive inhibition training. *Journal of Cognitive Neuroscience, 12*(5), 721–728. doi: 10.1162/089892900562525

Howlett, M. (2009). Governance modes, policy regimes and operational plans: A multi-level nested model of policy instrument choice and policy design. *Policy Sciences, 42*(1), 73–89. doi: 0.1007/s11077-009-9079-1

Howlett, M. (2014). Why are policy innovations rare and so often negative? Blame avoidance and problem denial in climate change policy-making. *Global Environmental Change, 29*. doi: 10.1016/j.gloenvcha.2013.12.009

Hughes, L. (2014). Climate change. In D. Lindenmayer, S. Dovers, & S. Morton (Eds.), *Ten commitments revisited: Securing Australia's future environment* (pp. 217–225). Collingwood, Australia: CSIRO Publishing.

Hulme, M. (2009). On the origin of 'the greenhouse effect': John Tyndall's 1859 interrogation of nature. *Weather, 64*(5), 121–123. doi: 10.1002/wea.386

Hulme, M. (2011). Reducing the future to climate: A story of climate determinism and reductionism. *Osiris, 26*(1), 245–266. doi: 10.1086/661274

Hulme, M. (2013). *Exploring climate change through science and in society: An anthology of Mike Hulme's essays*. London, UK: Routledge.

Hume, D. ([1739–1740] 2003). *A treatise of human nature*. New York, NY: Dover Publications.

Hupe, P. L., & Hill, M. J. (2015). 'And the rest is implementation'. Comparing approaches to what happens in policy processes beyond great expectations. *Public Policy and Administration, 5*(109), 1–19. doi: 0.1177/0952076715598828

Huskinson, L., & Stein, M. (2015). *Analytical psychology in a changing world: The search for self, identity and community*. New York, NY: Routledge.

Incropera, F. P. (2015). *Climate change: A wicked problem: Complexity and uncertainty at the intersection of science, economics, politics, and human behavior*. New York, NY: Cambridge University Press.

Innes, R., Gilchrist, L., Friedman, S., & Tompkins, K. (2016). John Dewey, participatory democracy, and university-community partnerships. In S. L. Barnes, L. Brinkley-Rubinstein, B. Doykas, N. C. Martin, & A. McGuire (Eds.), *Academics in action! A model for community-engaged research, teaching, and service* (pp. 27–48). New York, NY: Fordham University Press.

Institute of Medicine. (2013). *Environmental decisions in the face of uncertainty*. Washington, DC: National Academies Press.

IPCC. (1990). *Climate change: The IPCC scientific assessment (1990): Contribution of working group 1 to the first assessment report (FAR) of the Intergovernmental Panel on Climate Change*. New York, NY: Cambridge University Press.

IPCC. (2012). *Managing the risks of extreme events and disasters to advance climate change adaptation. A special report of working groups I and II of the Intergovernmental Panel on Climate Change*. New York, NY: Cambridge University Press.

IPCC. (2013). *Climate change 2013: The physical science basis. Contribution of working group I to the fifth assessment report of the Intergovernmental Panel on Climate Change* (Vol. 2013). New York, NY: Cambridge University Press.

IPCC. (2014a). *Climate change 2014: Mitigation of climate change. Contribution of working group III to the fifth assessment report of the Intergovernmental Panel on Climate Change*. New York, NY: Cambridge University Press.

IPCC. (2014b). *Climate change 2014: Synthesis report. Contribution of working groups I, II and III to the fifth assessment report of the Intergovernmental Panel on Climate Change*. New York, NY: Cambridge University Press.

IPCC. (2014c). Summary for policymakers. In O. Edenhofer, R. Pichs-Madruga, Y. Sokona, E. Farahani, K. Susanne, K. Seyboth, A. Adler, I. Baum, S. Brunner, P. Eickemeier, B. Kriemann, J. Savolainen, S. Schlömer, C. von Stechow, T. Zwickel, & J. C. Minx (Eds.), *Climate change 2014, mitigation of climate change. Contribution of working group III to the fifth assessment report of the Intergovernmental Panel on Climate Change*. New York, NY: Cambridge University Press.

IPCC. (2014d). Summary for policymakers. In R. K. Pachauri & L. Meyer (Eds.), *IPCC Fifth Assessment Synthesis Report. Contribution of Working Groups I, II, and III to the Fifth Assessment Report of the Intergovernmental Panel on Climate Change*. New York, NY: Cambridge University Press.

IPCC. (2014e). Summary for policymakers. In C. B. Field, V. R. Barros, D. J. Dokken, K. J. Mach, M. D. Mastrandrea, T. E. Bilir, M. Chatterjee, K. L. Ebi, Y. O. Estrada, R. C. Genova, B. Girma, E. S. Kissel, A. N. Levy, S. MacCracken, P. R. Mastrandrea, & L. L. White (Eds.), *Climate change 2014: Impacts, adaptation, and vulnerability. Part A: Global and sectoral aspects. Contribution of working group II to the fifth assessment report of the Intergovernmental Panel on Climate Change*. New York, NY: Cambridge University Press.

Jamieson, D. (2014). *Reason in a dark time: Why the struggle against climate change failed – and what it means for our future*. New York, NY: Oxford University Press.

Jamieson, K. H. (2017). The need for a science of science communication: Communicating science's values and norms. In K. H. Jamieson, D. Kahan, & D. A. Scheufele (Eds.), *The Oxford handbook of the science of science communication* (pp. 15–24). New York, NY: Oxford University Press.

Janoff-Bulman, R. (1992). *Shattered assumptions: Towards a new psychology of trauma*. New York, NY: Simon and Schuster.

Jenkins, T. (2013). *Of flying saucers and social scientists: A re-reading of when prophecy fails and of cognitive dissonance*. New York, NY: Palgrave Macmillan.

Johnson-Laird, P. N. (1983). *Mental models: Towards a cognitive science of language, inference, and consciousness*. Cambridge, MA: Harvard University Press.

Johnson-Laird, P. N. (2006). *How we reason*. New York, NY: Oxford University Press.

Johnson-Laird, P. N. (2013). *Human and machine thinking*. Hoboken, NJ: Taylor and Francis.

Johnson-Laird, P. N., & Byrne, R. M. J. (1991). *Deduction: Essays in cognitive psychology*. Hillsdale, NJ: Lawrence Erlbaum Associates.

Johnson-Laird, P. N., Khemlani, S. S., & Goodwin, G. P. (2015). Logic, probability, and human reasoning. *Trends in Cognitive Sciences, 19*(4), 201–214. doi: 10.1016/j.tics.2015.02.006

Johnston, M. (2014). *Corruption, contention, and reform: The power of deep democratization*. New York, NY: Cambridge University Press.

Jones, B. O. (2012). New media: Political infantilisation and the creativity paradox. In H. Sykes (Ed.), *More or less: Democracy and new media* (pp. 5–23). Albert Park, Australia: Future Leaders.

Jones, R. G. (2015). *Psychology of sustainability: An applied perspective*. New York, NY: Routledge.

Jordan, A., & O'Riordan, T. (1999). The precautionary principle in contemporary environmental policy and politics. In C. Raffensperger & J. A. Tickner (Eds.), *Protecting public health and the environment: Implementing the precautionary principle* (pp. 15–35). Washington, DC: Island Press.

Kagan, J. (1965). Individual differences in the resolution of response uncertainty. *Journal of Personality and Social Psychology, 2*(2), 154–160. doi: 10.1037/h0022199

Kagan, J., Rosman, B. L., Day, D., Albert, J., & Phillips, W. (1964). Information processing in the child: Significance of analytic and reflective attitudes. *Psychological Monographs: General and Applied, 78*(1), 1–37.

Kahan, D. M. (2010). Fixing the communications failure. *Nature, 463*(7279), 296. doi: 10.1038/463296a

Kahan, D. M. (2013). Ideology, motivated reasoning, and cognitive reflection. *Judgement and Decision Making, 8*(4), 407–424. doi: 10.2139/ssrn.2182588

Kahneman, D. (2003). A perspective on judgment and choice: Mapping bounded rationality. *The American Psychologist, 58*(9), 697–720. doi: 10.1037/0003-066X.58.9.697

Kahneman, D. (2011). *Thinking, fast and slow*. New York, NY: Farrar, Straus and Giroux.

Kahneman, D., & Tversky, A. (1973). On the psychology of prediction. *Psychological Review, 80*(4), 237–251. doi: 10.1037/h0034747

Kahneman, D., & Tversky, A. (1984). Choices, values, and frames. *American Psychologist, 39*(4), 341–350. doi: 10.1037/0003-066X.39.4.341

Kaku, M. (2014). *The future of the mind: The scientific quest to understand, enhance, and empower the mind*. New York, NY: Doubleday.

Kaplan, R., & Saccuzzo, D. (2012). *Psychological testing: Principles, applications, and issues*. Belmont, CA: Cengage Learning.

Keane, J. (2013). *Democracy and media decadence*. New York, NY: Cambridge University Press.

Keats, J. (2002). *Selected letters of John Keats*. Cambridge, MA: Harvard University Press.

Kee, C. (2017). *Not your average zombie: Rehumanizing the undead from voodoo to zombie walks*. Austin, TX: University of Texas Press.

Keynes, J. M. (1936). *The general theory of employment, interest and money*. London, UK: Macmillan.

Kim, H. S., & Sasaki, J. Y. (2013). Cultural neuroscience: Biology of the mind in cultural contexts. *Annual Review of Psychology, 65*(1), 487–514. doi: 10.1146/annurev-psych-010213-115040

Kingsolver, B. (2015). How to be hopeful. In S. Nicholson & P. K. Wapner (Eds.), *Global environmental politics: From person to planet* (pp. 339–344). Boulder, CO: Paradigm.

Kinley, R. (2016). Climate change after Paris: From turning point to transition. *Climate Policy, 17*(1), 9–15. doi: 10.1080/14693062.2016.1191009

Kitayama, S., & Park, J. (2010). Cultural neuroscience of the self: Understanding the social grounding of the brain. *Social Cognitive and Affective Neuroscience, 5*(2–3), 111–129. doi: 10.1093/scan/nsq052

Klabbers, J. (2008). Checks and balances in the law of international organisations In M. Sellers (Ed.), *Autonomy in the law*, pp. 141–164. Dordrecht, The Netherlands: Springer.

Klayman, J. (1988). Cue discovery in probabilistic environments: Uncertainty and experimentation. *Journal of Experimental Psychology: Learning, Memory, and Cognition, 14*(2), 317–330. doi: 10.1037/0278–7393.14.2.317

Komlos, J. (2014). *What every economics student needs to know and doesn't get in the usual principles text*. New York, NY: M. E. Sharpe.

Kopp, R. E., Simons, F. J., Mitrovica, J. X., Maloof, A. C., & Oppenheimer, M. (2013). A probabilistic assessment of sea level variations within the last interglacial stage. *Geophysical Journal International, 193*(2), 711–716. doi: 10.1093/gji/ggt029

Kosslyn, S. M. (1994). *Image and brain: The resolution of the imagery debate*. Cambridge, MA: MIT Press.

Kosslyn, S. M. (2005). Mental images and the brain. *Cognitive Neuropsychology, 22*(3), 333–347. doi: 10.1080/02643290442000130

Kouppanou, A. (2014). Imagining imagination and Bildung in the age of the digitized world picture. In M. Papastephanou, T. Strand, & A. Pirrie (Eds.), *Philosophy as a lived experience: Navigating through dichotomies of thought and action* (Vol. 3, pp. 77–89). Zürich, Switzerland: LIT Verlag GmbH and Co. KG Wien.

Kramer, R. (2014). Climate change: A state-corporate crime perspective. In T. Spapens, R. White, & M. Kluin (Eds.), *Environmental crime and its victims: Perspectives within green criminology*. Surrey, UK: Ashgate.

Krech, D., & Crutchfield, R. S. (1948). *Theory and problems of social psychology*. New York, NY: McGraw-Hill.

Kruglanski, A. W. (1989). *Lay epistemics and human knowledge: Cognitive and motivational bases*. New York, NY: Plenum Press.

Kuhn, T. S. ([1962] 2012). *The structure of scientific revolutions: 50th anniversary edition*. London, UK: University of Chicago Press.

Kunisue, Y. (2014). Listening to the natural world: Ecopsychology of listening from a Hawai'ian spiritual perspective. In J. A. Peeples & S. P. Depoe (Eds.), *Voice and environmental communication* (pp. 228–240). New York, NY: Palgrave Macmillan.

Kurzban, R., Burton-Chellew, M. N., & West, S. A. (2015). The evolution of altruism in humans. *Annual Review of Psychology, 66*, 575–599. doi: 10.1146/annurev-psych-010814-015355

Kurzweil, R. (2005). *The singularity is near: When humans transcend biology*. New York, NY: Viking.

Kuss, D. J. (2015). Treating Internet addiction – the expert perspective. *European Psychiatry, 30*(1), 464–464. doi: 10.1016/S0924-9338(15)30368-0

Kuss, D. J., & Griffiths, M. D. (2015). *Internet addiction in psychotherapy*. Hampshire, UK: Palgrave Macmillan.

Lakoff, G., & Johnson, M. (2008). *Metaphors we live by*. Chicago, IL: University of Chicago Press.

Lane, E. R. (1997). Friendship or commodities? The road not taken: Friendship, consumerism and happiness. In N. Goodwin, F. Ackerman, & D. Kiron (Eds.), *The consumer society* (pp. 101–103). Washington, DC: Island Press.

Langer, E. J., Blank, A., & Chanowitz, B. (1978). The mindlessness of ostensibly thoughtful action: The role of 'placebic' information in interpersonal interaction. *Journal of Personality and Social Psychology, 36*(6), 635–642. doi: 10.1037/0022-3514.36.6.635

Lasswell, H. D. (1948). *The analysis of political behaviour*. London, UK: Routledge.

Lasswell, H. D. (1960). *Psychopathology and politics*. New York, NY: The Viking Press.

Latif, M. (2011). Uncertainty in climate change projections. *Journal of Geochemical Exploration, 110*(1), 1–7. doi: 10.1016/j.gexplo.2010.09.011

Lau, R. R. (1982). Negativity in political perception. *Political Behavior, 4*(4), 353–377. doi: 10.1007/bf00986969

Lebow, R. N. (2018). *The politics and business of self-interest from Tocqueville to Trump*. London, UK: Palgrave Macmillan.

Lee, T. M., Markowitz, E. M., Howe, P. D., Ko, C., & Leiserowitz, A. A. (2015). Predictors of public climate change awareness and risk perception around the world. *Nature Climate Change, 5*(11), 1014–1020. doi: 10.1038/nclimate2728

Lee, J., & Xu, W. (2017). The more attacks, the more retweets: Trump's and Clinton's agenda setting on Twitter. *Public Relations Review*, doi: 10.1016/j.pubrev.2017.10.002

Lei, G., & Vargo, C. (2017). Global intermedia agenda setting: A big data analysis of international news. *Journal of Communication, 67*(4), 499–520. doi: 10.1111/jcom.12311

Leiserowitz, A. A. (2004). Before and after the day after tomorrow: A US study of climate change risk perception. *Environment, 46*(9), 22–37.

Lempert, R. J. (2002). A new decision sciences for complex systems. *Proceedings of the National Academy of Sciences of the United States of America, 99 Supplement, 3*(3), 7309–7313. doi: 10.1073/pnas.082081699

Lempert, R. J. (2015). Embedding (some) benefit-cost concepts into decision support processes with deep uncertainty. *Journal of Benefit-Cost Analysis, 5*(3), 487–514. doi: 10.1515/jbca-2014-9006

Lempert, R. J., & Groves, D. G. (2010). Identifying and evaluating robust adaptive policy responses to climate change for water management agencies in the American west. *Technological Forecasting and Social Change, 77*(6), 960–974. doi: 10.1016/j.techfore.2010.04.007

Lempert, R. J., Groves, D. G., Popper, S. W., & Bankes, S. C. (2006). A general, analytic method for generating robust strategies and narrative scenarios. *Management Science, 52*(4), 7309–7313. doi: 10.1287/mnsc.1050.0472

Lesch, K. P., & Canli, T. (2007). Long story short: The serotonin transporter in emotion regulation and social cognition. *Nature Neuroscience, 10*(9), 1103–1109. doi: 10.1038/nn1964

Lester, P. M. (2013). *Visual communication: Images with messages* (6th edn). Boston, MA: Wadsworth, Cengage Learning.

Levin, J. (2016). Consciousness. In B. P. McLaughlin (Ed.), *Philosophy: Mind* (pp. 119–143). New York, NY: Macmillan.

Levin, K., Cashore, B., Bernstein, S., & Auld, G. (2012). Overcoming the tragedy of super wicked problems: Constraining our future selves to ameliorate global climate change. *Policy Sciences, 45*(2), 123–152. doi: 10.1007/s11077-012-9151-0

Leviston, Z., Greenhill, M., & Walker, I. (2015). *Australian attitudes to climate change: 2010–2014*. Clayton, Australia. CSIRO Publishing. Retrieved 15 July, 2016, from https://publications.csiro.au/rpr/pub?list=SEA&pid=csiro:EP158 008.

Lieberman, D. A. (2016). *The case against free will: What a quiet revolution in psychology has revealed about how behavior is determined*. New York, NY: Palgrave Macmillan.

Lindblom, C. E. (1959). The science of 'muddling through'. *Public Administration Review, 19*(2), 79–88. doi: 10.2307/973677

Lindblom, C. E., & Woodhouse, E. J. (1992). *The policy-making process* (3rd edn). Englewood Cliffs, NJ: Prentice Hall.

Lipari, L. (2014). *Listening, thinking, being: Toward an ethics of attunement*. University Park, PA: Pennsylvania State University Press.

Lippmann, W. (1922). *Public opinion*. London, UK: Allen and Unwin.

Lippmann, W. (1927). *The phantom public*. New Brunswick, NJ: Transaction.

Liu, J., & Scheufele (2016). A revisionist perspective on framing effects. *Politics*, doi: 10.1093/acrefore/9780190228637.013.48

Lloyd, E., & Schweizer, V. (2014). Objectivity and a comparison of methodological scenario approaches for climate change research. *Synthese, 191*(10), 2049–2088. doi: 10.1007/s11229-013-0353-6

Lobel, T. (2014). *Sensation: The new science of physical intelligence*. New York, NY: Atria Books.

Lomborg, B. (2015). Impact of current climate proposals. *Global Policy, 7*(1), 109–118. doi: 10.1111/1758-5899.12295

Lorenz, E. (1972). *Predictability: Does the flap of a butterfly's wings in Brazil set off a tornado in Texas*. Paper presented at the 139th Meeting of the American Association for the Advancement of Science, Washington, DC. http://web. mit.edu/lorenzcenter/about/LorenzPubs/Butterfly_1972.pdf.

Louw, E. (2010). *The media and political process* (2nd edn). Washington, DC: Sage.

Lovejoy, A. O. ([1936] 1964). *The great chain of being*. Cambridge, MA: Harvard University Press.

Lovelock, J. (2007a). *The revenge of Gaia: Earth's climate in crisis and the fate of humanity*. New York, NY: Penguin.

Lovelock, J. (2007b). *The revenge of Gaia: Why the Earth is fighting back and how we can still save humanity*. New York, NY: Penguin.

Luntz, F. (2002). *The environment: A cleaner, safer, healthier America*. Alexandria, VA: The Luntz Research Companies – Straight Talk. Retrieved 12 March, 2015, from www2.bc.edu/~plater/Newpublicsite06/suppmats /02.6.pdf.

MacLean, N. (2017). *Democracy in chains: The deep history of the radical right's stealth plan for America*. New York, NY: Viking.

Macnamara, B. N., Hambrick, D. Z., & Oswald, F. L. (2014). Deliberate practice and performance in music, games, sports, education, and professions: A meta-analysis. *Psychological Science, 25*(8), 1608–1618. doi: 10.1177/0956797614535810

Madison, J. (1788). *The federalist No. 51*. In The federalist. New York, NY: Modern Library.

Maddison, S., & Denniss, R. (2013). *An introduction to Australian public policy: Theory and practice* (2nd edn). New York, NY: Cambridge University Press.

Magistretti, P. J. (2015). A cellular perspective on brain energy metabolism and functional imaging. *Neuron, 86*(4), 883–901. doi: 10.1016/j.neuron.2015.03.035

Maguire, E. A., Gadian, D. G., Johnsrude, I. S., Good, C. D., Ashburner, J., Frackowiak, R. S., & Frith, C. D. (2000). Navigation-related structural change in the hippocampi of taxi drivers. *Proceedings of the National Academy of Sciences of the United States of America, 97*(8), 4398–4403. doi: 10.1073/pnas.070039597

Malinen, K., & Matta, C. F. (2018). 'Climate change' and the 'butterfly effect' in an eighteenth century monograph. *Foundations of Chemistry, 1*(1), 1–4. doi: 10.1007/s10698-018-9305-z

Mann, T. (2013). *Australian law dictionary* (2nd edn). South Melbourne, Australia: Oxford University Press.

Markovits, H. (2014). Introduction. In H. Markovits (Ed.), *The developmental psychology of reasoning and decision-making* (pp. 1–6). East Sussex, UK: Psychology Press.

Markovits, H., & Barrouillet, P. (2002). The development of conditional reasoning: A mental model account. *Developmental Review, 22*(1), 5–36. doi: 10.1006/drev.2000.0533

Matsubayashi, T. (2013). Do politicians shape public opinion? *British Journal of Political Science, 43*(2), 451–478. doi: 10.1017/S0007123412000373

Matthen, M. (2014). Active perception and the representation of space. In D. Stokes, M. Matthen, & S. Biggs (Eds.), *Perception and its modalities* (pp. 44–72). New York, NY: Oxford University Press.

Matthes, J. (2012). Framing politics: An integrative approach. *American Behavioral Scientist, 56*(3), 247–259. doi: 10.1177/0002764211426324

Mauss, M. (1990). *The gift: The form and reason for exchange in archaic societies* (W. D. Halls, Trans.). London, UK: Routledge.

Maxwell, R., & Miller, T. (2015). Greening media studies. In R. Maxwell, J. Raundalen, & N. A. L. Vestberg (Eds.), *Media and the ecological crisis* (pp. 87–98). New York, NY: Routledge.

May, C. (2013). Towards a general theory of implementation. *Implementation Science, 8*(1), 8–18. doi: 10.1186/1748-5908-8-18

May, P. J. (2012). Policy design and implementation. In B. G. Peters & J. Pierre (Eds.), *The Sage handbook of public administration* (2nd edn, pp. 279–291). London, UK: Sage.

Mayer, L. A., Loa, K., Cwik, B., Tuana, N., Keller, K., Gonnerman, C., Parker, A. M., & Lempert, R. J. (2017). Understanding scientists' computational modeling decisions about climate risk management strategies using values-informed mental models. *Global Environmental Change, 42*, 107–116. doi: 10.1016/j.gloenvcha.2016.12.007

Mayers, T. (2017). Forward. In A. Koehler, *Composition, creative writing studies, and the digital humanities* (pp. xi–xiv). London, UK: Bloomsbury.

Mazur, A. (2018). *Technical controversies over public policy: From fluoridation to fracking and climate change*. Cambridge, UK: Routledge.

McCarthy, J. & Hayes, P. J. (1969). Some philosophical problems from the standpoint of artificial intelligence. In B. Meltzer & D. Michie (Eds.), *Machine intelligence 4* (pp. 463–502). Edinburgh, UK: Edinburgh University Press.

McCombs, M. (2014). *Setting the agenda: The mass media and public opinion* (2nd edn). Cambridge, UK: Polity Press.

McCroskey, J. C., & Jenson, T. A. (1975). Image of mass media news sources. *Journal of Broadcasting, 19*(2), 169–180. doi: 10.1080/08838157509363777

McDowell, P. E. (2015). *Thinking about thinking: Cognition, science, and psychotherapy*. New York, NY Routledge.

McGrath, M. (2017, October 10.). Anger over 'untrue' climate change claims. *BBC News*. Retrieved 25 January, 2018, from www.bbc.com/news/science-environment-40899188.

McKibben, B. (2017, August 8.). The Trump administration's solution to climate change: Ban the word. *Guardian*. Retrieved 26 January, 2018, from www.theguardian.com/commentisfree/2017/aug/08/trump-administration-climate-change-ban-usda.

McKnight, D. (2012). *Rupert Murdoch: An investigation of political power*. Sydney, Australia: Allen & Unwin.

McLuhan, M. (2013). *Understanding media: The extensions of man*. New York, NY: Gingko.

McRaney, D. (2012). *You are not so smart: Why your memory is mostly fiction*. Oxford, UK: Penguin.

Medhaug, I., Stople, M. B., Fischer, E. M., & Knutti, R. (2017). Reconciling controversies about the 'global warming hiatus'. *Nature, 545*, 41–47. doi: 10.1038/nature22315

Meldrum, H., Szymanski, D., Oches, E. A., & Davis, P. T. (2017). Speaking out or staying quiet on climate change: Broadcast meteorologists influenced by the need to be pithy, popular and politically cautious. In W. L. Filho, & J. M. Keenan (Eds.), *Climate change adaptation in North America* (pp. 261–277). Cham, Switzerland: Springer.

Mesquita, B., Barrett, L. F., & Smith, E. R. (2010). *The mind in context*. New York, NY: Guilford Publications.

Meyer, J. W., & Rowan, B. (1977). Institutionalized organizations: Formal structure as myth and ceremony. *American Journal of Sociology, 83*(2), 340–363.

Meyer, T., & Hinchman, L. P. (2002). *Media democracy: How the media colonize politics*. Cambridge, UK: Polity Press.

Michaelian, K. (2016). Memory. In B. P. McLaughlin (Ed.), *Philosophy: Mind* (pp. 227–243). New York, NY: Macmillan.

Mikkelsen, T. B., Grinsted, A., & Ditlevsen, P. (2018). Influence of temperature fluctuations on equilibrium ice sheet volume. *Cryosphere, 12*, 39–47. doi: 10.5194/tc-12-39-2018

Mill, J. S. ([1848] 2001). *The principles of political economy*. Kitchener, Canada: Batoche Books.

Mill, J. S. ([1859] 1982). *On liberty*. London, UK: Penguin.

Miller, A. H. (1974). Political issues and trust in government: 1964–1970. *American Political Science Review, 68*(3), 951–972. doi: 10.2307/1959140

Mills, C. W. (1959). *The sociological imagination*. New York, NY: Oxford University Press.

Mills, K. L. (2014). Effects of Internet use on the adolescent brain: Despite popular claims, experimental evidence remains scarce. *Trends in Cognitive Sciences, 18*(8), 385–387. doi: 10.1016/j.tics.2014.04.011

Mole, C., & Zhao, J. (2015). Vision and abstraction: An empirical refutation of Nico Orlandi's non-cognitivism. *Philosophical Psychology, 29*(3–2016), 365–373.

Möller, F. (1963). On the influence of changes in the CO2 concentration in air on the radiation balance of the Earth's surface and on the climate. *Journal of Geophysical Research, 68*(13), 3877–3886.

Monopolis, A. (2012). *Infrastructure sustainability and design*. New York, NY: Taylor & Francis.

Mora, C., Frazier, A. G., Longman, R. J., Dacks, R. S., Walton, M. M., Tong, E. J., … Giambelluca, T. W. (2013). The projected timing of climate departure from recent variability. *Nature, 502*(7470), 183–187. doi: 10.1038/nature12540

Mortreux, C., & Barnett, J. (2009). Climate change, migration and adaptation in Funafuti, Tuvalu. *Global Environmental Change, 19*(1), 105–112. doi: 10.1016/j.gloenvcha.2008.09.006

Moser, S. (2016). Reflections on climate change communication research and practice in the second decade of the 21st century. *WIREs Climate Change, 7*(3), 345–369. doi: 10.1002/wwc.403

Moscovici, S. (1988). Notes towards a description of social representations. *European Journal of Social Psychology, 18*(3), 211–250. doi: 10.1002/ejsp.2420180303

Moshman, D. (2013). Epistemic cognition and development. In P. Barrouillet & C. Gauffroy (Eds.), *The development of thinking and reasoning* (pp. 13–33). London, UK: Psychology Press.

Moshman, D. (2014). Epistemic domains of reasoning. In H. Markovits (Ed.), *The developmental psychology of reasoning and decision-making* (pp. 13–33). East Sussex, UK: Psychology Press.

Mueller, D. C. (2003). *Public choice III*. New York, NY: Cambridge University Press.

Mungiu-Pippidi, A. (2015). *The quest for good governance: How societies develop control of corruption*. Cambridge, UK: Cambridge University Press.

Nagel, T. (1970). *The possibility of altruism*. Princeton, NJ: Princeton University Press.

Nair, S., & Howlett, M. (2017). The central conundrums of policy formulation: Illstructured problems and uncertainty. In M. Howlett, & I. Mukherjee (Eds.), *Handbook of policy formulation* (pp. 23–38). Northampton, MA: Edward Elgar.

National Aeronautics and Space Administration. (2018). *Long-term warming trend continued in 2017: NASA, NOAA.* Washington, DC: NASA. Retrieved 17 January, 2018, from www.nasa.gov/news/2671/long-term-warming-trend-continued-in-2017-nasa-noaa/.press-release/nasa-noaa-data-show-2016-warmest-year-on-record-globally.

Naznin, N., Hossain, R., & Huq, S. (2017). Post-Paris long-term climate capacity: The role of universities. In T. B. B. Scavenius & S. Rayner (Eds.), *Institutional capacity for climate change response: A new approach to climate politics*, (pp. 134–146). New York, NY: Taylor & Francis.

Neander, K. (2016). Mental content. In B. P. McLaughlin (Ed.), *Philosophy: Mind* (pp. 145–156). New York, NY: Macmillan.

Newell, B. R., & Shanks, D. R. (2014). Unconscious influences on decision making: A critical review. *The Behavioral and Brain Sciences, 37*(1), 1–1. doi: 10.1017/s0140525x12003214

Newsinger, J. (2018). *Hope lies in the proles*. London, UK: Pluto Press.

Nicholson, S., & Wapner, P. K. (2015). *Global environmental politics: From person to planet*. Boulder, CO: Paradigm.

Nickerson, R. S., Perkins, D. N., & Smith, E. E. (2014). *The teaching of thinking.* Hoboken, NJ: Taylor & Francis.

Niebert, K., & Gropengiesser, H. (2013). Understanding and communicating climate change in metaphors. *Environmental Education Research, 19*(3), 282–302. doi: 10.1080/13504622.2012.690855

Nietzsche, F. ([1884] 1974). *The gay science* (W. Kaufman, Trans.). New York, NY: Vintage Books.

Nisbet, M. C. (2009). Framing science: A new paradigm in public engagement. In L. Kahlor & P. A. Stout (Eds.). *Understanding Science: New Agendas in Science Communication* (pp. 40–67). New York, NY: Routledge.

Nisbett, R. E., Peng, K., Choi, I., & Norenzayan, A. (2001). Culture and systems of thought: Holistic versus analytic cognition. *Psychological Review, 108*(2), 291–310. doi: 10.1037/0033-295X.108.2.291

Nisbett, R. E., & Ross, L. (1980). *Human inference: Strategies and shortcomings of social judgment.* Englewood Cliffs, NJ: Prentice Hall.

Norman, G. R., Monteiro, S. D., Sherbino, J., Ilgen, J. S., Schmidt, H. G., & Manede, S. (2017). The causes of errors in clinical reasoning: Cognitive biases, knowledge deficits, and dual processing thinking. *Journal of the Association of American Medical Colleges, 92*(1), 23–30. doi: 10.1097/ACM.0000000000001421

Northrop, R. B., & Connor, A. N. (2013). *Ecological sustainability: Understanding complex issues*. Boca Raton, FL: CRC Press.

Nussbaum, M. C. (2012). *Not for profit: Why democracy needs the humanities*. Princeton, NJ: Princeton University Press.

O'Connor, K. P. (2015). *A constructionist clinical psychology for cognitive behaviour therapy*. New York, NY: Routledge.

O'Neill, J. ([1944] 2007). *Prodigal genius: The life of Nikola Tesla*. San Diego, CA: Book Tree.

Oakes, L. (2013). *Remarkable times: Australian politics 2010–13: What really happened.* Sydney, Australia: Hachette.

Oaksford, M., & Chater, N. (2001). The probabilistic approach to human reasoning. *Trends in Cognitive Sciences, 5*(8), 349–357. doi: 10.1016/S1364-6613(00)01699-5

Oaksford, M., & Chater, N. (2009). Précis of Bayesian Rationality: The probabilistic approach to human reasoning. *Behavioural and Brain Sciences, 32*(1), 69–84. doi: 10.1017/S0140525X09000284

Ober, J. (2008). *Democracy and knowledge: Innovation and learning in classical Athens*. Princeton, NJ: Princeton University Press.

OECD Convention on Combating Bribery of Foreign Public Officials in International Business Transactions. (2014). Ratification status as of 21 May 2014. Retrieved 23 May, 2016, from www.oecd.org/ daf/anti-bribery/WGB RatificationStatus.pdf.

Offord, B., Kerruish, E., Garbutt, R., Pavlovic, K., & Wessell, A. (2014). *Inside Australian culture: Legacies of enlightenment values*. New York, NY: Anthem Press.

Olson, M., & Fazio, R. H. (2012). Implicit and explicit measures of attitudes: The perspective of the MODE model. In R. E. Petty, R. H. Fazio, & P. Brinol (Eds.), *Attitudes: Insights from the new implicit measures* (pp. 19–63). New York, NY: Psychology Press.

Orlandi, N. (2014). *The innocent eye: Why vision is not a cognitive process*. New York, NY: Oxford University Press.

Ornstein, R. E., & Ehrlich, P. R. (2011). *New world new mind: Moving toward conscious evolution*. Los Altos, CA: ISHK.

Orr, D. W. (2014). Foreword. In T. W. Institute (Ed.), *State of the world 2014: Governing for sustainability*. Washington, DC: Island Press.

Orwell, G. (1949). *Nineteen eighty-four (1984)*. London, UK: Secker & Warburg.

Orwell, G. (1968). Politics and the English language. In S. Orwell & I. Angos (Eds.), *The collected essays, journalism and letters of George Orwell* (pp. 127–140). New York, NY: Harcourt, Brace, Javanovich.

Osbeck, L. M., Nersessian, N. J., Malone, K. R., & Newstetter, W. C. (2010). *Science as psychology: Sense-making and identity in science practice*. New York, NY: Cambridge University Press.

Östman, J. (2014). The influence of media use on environmental engagement: A political socialization approach. *Environmental Communication, 8*(1), 92–109. doi: 10.1080/17524032.2013.846271

Ourbak, T. (2017). Changing the game: Paris Agreement and the role of scientific communities. *Climate Policy, 17*(7), 819824. doi: 10.1080/14693062.2017.1348331

Oxfam. (2016). *An economy for the 1%: How privilege and power in the economy drive extreme inequality and how this can be stopped*. Retrieved 7 March, 2016, from http://oxfamilibrary.openrepository.com/oxfam/bitstream/10546/592643/47/bp210-economy-one-percent-tax-havens-180116-en.pdf.

Paffenroth, K. (2011). Zombies as internal fear or threat. In S. Boluk & W. Lenz (Eds.), *Generation zombie: Essays on the living dead in modern culture* (pp. 18–26). New York, NY: McFarland & Company.

Page, B. I., Shapiro, R. Y., Dempsey, G. R. (1987). What moves public opinion? *American Political Science Review, 81*(1), 23–43. doi: 10.2307/1960777

Painter, J. (2013). *Climate change in the media: Reporting risk and uncertainty*. New York, NY: Palgrave Macmillan.

Pal, J., & Gonawela, A. (2017). Studying political communication on Twitter: The case for small data. *Current Opinion in Behavioral Sciences, 18*, 97–102. doi: 10.1016/j.cobeha.2017.09.009

Papathanassopoulos, S., Coen, S., Curran, J., Aalberg, T., Rowe, D., Jones, P., ... Tiffen, R. (2013). Online threat, but television is still dominant: A comparative study of 11 nations' news consumption. *Journalism Practice, 7*(6), 690–704. doi: 10.1080/17512786.2012.761324

Parsons, T. (1990). Prolegomena to a theory of social institutions. *American Sociological Review, 55*(3), 319–333.

Patterson, J. J., Thaler, T., Hoffman, M., Hughes, S., Oels, A., Chu, E., ... Jordan, A. (2018). Political feasibility of 1.5°C societal transformations: The role of social justice. *Environmental Sustainability, 31*, 1–9. doi: 10.1016/j.cosus.2017.11.002

Peck, J. (2010). Zombie neoliberalism and the ambidextrous state. *Theoretical Criminology, 14*(1), 104–110. doi: 10.1177/1362480609352784

Pennycook, A. (2017). *The cultural politics of English as an international language.* New York, NY: Routledge.

Pennycook, G., Cheyne, J. A., Seli, P., Koehler, D. J., & Fugelsang, J. A. (2012). Analytic cognitive style predicts religious and paranormal belief. *Cognition, 123*(3), 335–346. doi: 10.1016/j.cognition.2012.03.003

Pepper, S. C. ([1942] 1970). *World hypotheses: A study in evidence.* Berkley, CA: University of California Press.

Petty, R. E., & Cacioppo, J. T. (1986). The elaboration likelihood of persuasion. In L. Berowitz (Ed.), *Advances in Experimental Social Psychology* (Vol. 19, pp. 123–205). New York, NY: Academic Press.

Pew Research Center. (2014). *State of the news media.* Retrieved 26 February, 2017, from http://assets.pewresearch.org/wp-content/uploads/sites/13/2017/05/30142556/state-of-the-news-media-report-2014-final.pdf.

Pew Research Center. (2017). *Public trust in government: 1958–2017.* Retrieved 26 July, 2017, from www.people-press.org/2017/05/03/public-trust-in-government-1958–2017/.

Pezzey, J. C. V. (2014). The influence of lobbying on climate policies; or, why the world might fail. *Environment and Development Economics, 19*(03), 329–332. doi: 10.1017/S1355770X14000345

Philp, M. (2015). The definition of political corruption. In P. M. Heywood (Ed.), *Routledge handbook of political corruption* (pp. 17–29). New York, NY: Routledge.

Pidgeon, N., & Fischhoff, B. (2014). The role of social and decision sciences in communicating uncertain climate risks. In J. Arvai & L. Rivers (Eds.), *Effective risk communication* (pp. 319–332). London, UK: Routledge.

Pietersen, H. J. (2014). Science and religion in meta-perspective – Part I. *Mediterranean Journal of Social Sciences, 5*(20), 1969. doi: 10.5901/mjss.2014.v5n20p1969

Pirie, M. (2009). *101 great philosophers.* New York, NY: Continuum International Publishing.

Polanyi, K. ([1944] 2001). *The great transformation: The political and economic origins of our time* (2nd edn). Boston, MA: Beacon Press.

Polanyi, M. ([1958] 2005). *Personal knowledge: Towards a post-critical philosophy.* London, UK: Routledge.

Pompa, L. ([1725] 2002). *The first new science.* Cambridge, UK: Cambridge University Press.

Pontell, H. N., Black, W. K., & Geis, G. (2014). Too big to fail, too powerful to jail? On the absence of criminal prosecutions after the 2008 financial meltdown. *Crime, Law and Social Change, 61*(1), 1–13. doi: 10.1007/s10611-013-9476-4

Pooley, E. (2017). Climate change denial is the original fake news. *Time.* Retrieved 26 January, 2018, from http://time.com/4664173/climate-change-denial-fake-news/.

Popper, K. R. (1959). *The logic of scientific discovery*. London, UK: Hutchinson.

Popper, K. R. (1963). *Conjectures and refutations: The growth of scientific knowledge*. New York, NY: Routledge.

Pornpattananangkul, N., Cheon, B. K., & Chiao, J. Y. (2014). The role of negativity bias in political judgment: A cultural neuroscience perspective. *The Behavioral and Brain Sciences, 37*(3), 325–326. doi: 10.1017/s0140525x13002707

Poteete, A. R., Janssen, M. A., & Ostrom, E. (2010). *Working together: Collective action, the commons, and multiple methods in practice*. Princeton, NJ: Princeton University Press.

Pressman, J. L., & Wildavsky, A. B. (1973). *Implementation: How great expectations in Washington are dashed in Oakland or why it's amazing that Federal programs work at all*. Berkley, CA: University of California Press.

Pribram, K. H. (2011). *Brain and perception: Holonomy and structure in figural processing*. New York, NY: Routledge.

Princen, S. (2018). Agenda-setting and framing in Europe. In E. Ongaro & S. Van Thiel (Eds.), *Handbook of public administration and management in Europe* (pp. 535–551). London, UK: Palgrave Macmillan.

Pruitt, S. (2017). *Senate environment and public works committee hearing, 'Nomination of attorney general Scott Pruitt to be administrator of the U.S. environmental protection agency'*. Retrieved 27 January, 2018, from www.epw.senate.gov/public/_cache/files/6d95005c-bd1a-4779-af7e-be831db6866a/scott-pruitt-qfr-responses-01.18.2017.pdf.

Pylyshyn, Z. W. (1981). The imagery debate: Analogue media versus tacit knowledge. *Psychological Review, 88*(1), 16–45. doi: 10.1037/0033-295X.88.1.16

Pylyshyn, Z. W. (2002). Mental imagery: In search of a theory. *Behavioral and Brain Sciences, 25*(2), 157–182. doi: 10.1017/S0140525X02000043

Quiquet, A., Ritz, C., Punge, H. J., & Melia, D. S. Y. (2013). Greenland ice sheet contribution to sea level rise during the last interglacial period: A modelling study driven and constrained by ice core data. *Climate of the Past, 9*(1), 353–366. doi: 10.5194/cp-9-353-2013

Rahmstorf, S. (2001). *Encyclopedia of ocean sciences* (pp. 1–7). Waltham, MA: Academic Press.

Rahmstorf, S. (2009). *The Day After Tomorrow* – some comments on the movie. Retrieved 27 May, 2014, from www.pik-potsdam.de/~tefan/tdat_review.html.

Ramm, T. D., Watson, C. S., & White, C. J. (2018). Describing tipping points in coastal flood risk management. *Computers, Environment and Urban Systems*. doi: 10.1016/j.compenvurbsys.2018.01.002

Ravetz, J. R. (2011). 'Climategate' and the maturing of post-normal science. *Futures, 43*(2), 149–157. doi: 10.1016/j.futures.2010.10.003

Rawls, J. (1971). *A theory of justice* (Revised edition). Cambridge, MA: Harvard University Press.

Rawls, J. (2011). *Political liberalism: Expanded edition* (Vol. 1). Princeton, NJ: Princeton University Press.

Rayner, T. (2018). Feeling the heat: The challenge of communicating 'high-end' climate change. In F. W. Leal, E. Manolas, A. Azul, U. Azeiteiro, H. McGhie (Eds.) *Handbook of climate change communication: Vol. 1. Climate change management* (pp. 367–383). Cham, Switzerland: Springer.

Reber, A. S. (1989). Implicit learning and tacit knowledge. *Journal of Experimental Psychology: General, 118*(3), 219–235. doi: 10.1037/0096-3445.118.3.219

Reese, R., & Wernick, P. (2009). *Zombie land*. Culver City, CA: Sony Pictures.

Reisberg, D. (2013). Mental images. In D. Reisberg (Ed.), *The Oxford handbook of cognitive psychology*. Oxford, UK: Oxford University Press.

Rhodes, G. D. (2001). *White zombie: Anatomy of a horror film.* Jefferson, NC: McFarland.

Richardson, J. T. E. (2013). *Imagery*. Hoboken, NJ: Taylor & Francis.

Richtel, M. (2014). *A deadly wandering: A tale of tragedy and redemption in the age of attention*. New York, NY: HarperCollins.

Richter, D. (2014). *Historical dictionary of Wittgenstein's philosophy*. New York, NY: Rowman & Littlefield.

Rigby, S. H. (1987). *Marxism and history: A critical introduction*. New York, NY: St. Martin's Press.

Riggs, A., & Turner, B. S. (2000). Pie-eyed optimists: Baby-boomers the optimistic generation? *Social Indicators Research, 52*(1), 73–93. doi: 10.1023/a:1007079807500

Rio Declaration on Environment and Development. (1992). *Rio Declaration on Environment and Development*. Stockholm, Sweden: United Nations. Retrieved 5 March, 2017, from www.unep.org/Documents.Multilingual/Default. Print.asp?documentid=78&articleid=1163.

Roberts, E., & Pelling, M. (2018). Climate change-related loss and damage: Translating the global policy agenda for national policy processes. *Climate and Development, 10*(1), 4–17. doi: 10.1080/17565529.2016.1184608

Robinson, R. (2004). Neoliberalism and the future world: Markets and the end of politics. *Critical Asian Studies, 36*(3), 405–423. doi: 10.1080/1467271042000241603

Roese, N. J. (1997). Counterfactual thinking. *Psychological Bulletin, 121*(1), 133–148. doi: 10.1037/0033-2909.121.1.133

Rogoff, B. (2003). *The cultural nature of human development*. Oxford, UK: Oxford University Press.

Rohde, R. A. & Muller, R. A. (2015). Air pollution in China: Mapping of concentrations and sources. *PLOS ONE, 10*(8), 1–14. doi:10.1371/journal.pone.0135749

Rorty, R. (1979). *Philosophy and the mirror of nature*. Princeton, NJ: Princeton University Press.

Rosen, J. (2017). Political turbulence ahead. *Nature, 544*, 509–511. doi: 10.1038/nj7651–509a

Rosenzweig, R. H. (2016). *Global climate change policy and carbon markets*. London, UK: Palgrave Macmillan.

Ross, J. J. (2009). *Reading Wittgenstein's philosophical investigations: A beginner's guide*. Plymouth, UK: Lexington.

Roth, W. M., & Bowen, G. M. (2003). When are graphs worth ten thousand words? An expert-expert study. *Cognition and Instruction, 21*(4), 429–473. doi: 10.1207/s1532690xci2104_3

Rousseau, J. ([1762] 2003). *On the social contract*. New York, NY: Dover Publications.

Royal Commission into Institutional Responses to Child Sexual Abuse. (2016). *Officers of the Royal Commission*. Canberra, Australia: Australian Government. Retrieved 9 November, 2017, from www.childabuseroyalcommission.gov.au/about-us/terms-of-reference.

Ruckelshaus, W. D. (1992). Toward a sustainable world. In J. Hollander (Ed.), *The energy-environment connection* (pp. 365–377). Washington, DC: Island Press.

Ruda, F. (2014). *For Badiou: Idealism without idealism*. Evanston, IL: Northwestern University Press.

Rupert, R. D. (2009). *Cognitive systems and the extended mind*. Oxford, UK: Oxford University Press.

Ruse, M. (2004). *Can a Darwinian be a Christian? The relationship between science and religion*. Cambridge, UK: Cambridge University Press.

Ruse, M. (2007). My journey in the world of religion-and-science. *Zygon, 42*(3), 577–582. doi: 10.1111/j.1467-9744.2007.00849.x

Russell, B. ([1912] 2001). *The problems of philosophy*Oxford, UK: Oxford University Press.

Russell, M. ([1926] 2014). Parliamentary party cohesion: Some explanations from psychology. *Party Politics, 20*(5), 712–723. doi: 10.1177/1354068812453367

Rybak, J. P. (1999, November). Nikola Tesla: Scientific savant. *Popular Electronics, 16*, 40–48.

Ryle, G. (1949). *The concept of mind*. Chicago, IL: University of Chicago Press.

Salama, S. & Aboukoura, K. (2018). Role of emotions in climate change communication. In F. W. Leal, E. Manolas, A. Azul, U. Azeiteiro, & H. McGhie (Eds.), *Handbook of climate change communication: Vol. 1. Climate change management* (pp. 137–150). Cham, Switzerland: Springer.

Samuels, A. (2015). Making a difference? When individuals take personal responsibility for social and political change. In L. Huskinson & M. Stein (Eds.), *Analytical psychology in a changing world: The search for self, identity and community* (pp. 42–52). New York, NY: Routledge.

Sandel, M. J. (1998). *Liberalism and the limits of justice*. New York, NY: Cambridge University Press.

Sandfort, J., & Moulton, S. (2015). *Effective implementation in practice: Integrating public policy and* management (1st edn). San Francisco, CA: Jossey-Bass.

Santer, B. D., Solomon, S., Wentz, F. J., Fu, Q., Po-Chedley, S., Mears, C., … (2017). Tropospheric warming over the past two decades. *Nature Scientific Reports, 7*(2336), 617–636. doi: 10.1038/s41598-017-02520-7

Santayana, G. (1906). *The life of reason*. New York, NY: Charles Scribner's Sons.

Sartre, J. ([1936] 2012). *The imagination* (K. Williford & D. Rudrauf, Trans.). New York, NY: Routledge.

Saul, B., Sherwood, S., McAdam, J., Stephens, T., & Slezak, J. (2012). *Climate change and Australia: Warming to the global challenge*. Annandale, Australia: The Federation Press.

Saumier, D., & Chertkow, H. (2002). Semantic memory. *Current Neurology and Neuroscience Reports, 2*(6), 516–522. doi: 10.1007/s11910-002-0039-9

Schacter, D. L. (2001). *The seven sins of memory: How the mind forgets and remembers*. New York, NY: Houghton Mifflin.

Schäfer, M. (2011). Science journalism and fact checking. *Journal of Science Communication, 10*(4), 1–5.

Scheffran, J. (2015). Complexity and stability in human-environment interaction. In E. Kavalski (Ed.), *World politics at the edge of chaos: Reflections on complexity and global life* (pp. 229–254). Albany, NY: State University of New York.

Schenk, T. (2018). *Adapting infrastructure to climate ch*ange: Advancing decision-making under conditions of uncertainty. New York, NY: Routledge.

Scheufele, D. A. (1999). Framing as a theory of media effects. *Journal of Communication, 49*(1), 103–122. doi: 10.1111/j.1460-2466.1999.tb02784.x

Scheufele, D. A., & Tewksbury, D. (2006). Framing, agenda setting, and priming: The evolution of three media effects models. *Journal of Communication, 57*(1), 103–122. doi: 10.1111/j.0021-9916.2007.00326.x

Schlichting, I. (2013). Strategic framing of climate change by industry actors: A meta-analysis. *Environmental Communication, 7*(4), 493–511. doi: 10.1080/17524032.2013.812974

Schmitt, F. F. (2014). *Hume's epistemology in the treatise: A veritistic interpretation.* Oxford, UK: Oxford University Press.

Schnaiberg, A. (2002). Reflections on my 25 years before the mast of the environment and technology section. *Organization and Environment, 15*(1), 30–41.

Schneider, B. (2012). Climate model simulation visualization from a visual studies perspective. *Wiley Interdisciplinary Reviews: Climate Change, 3*(2), 185–193. doi: 10.1002/wcc.162

Schneider, W. (2015). *Memory development from early childhood through emerging adulthood.* Cham, Switzerland: Springer.

Schneider, W., & Shiffrin, R. M. (1977). Controlled and automatic human information processing: I. Detection, search, and attention. *Psychological Review, 84*(1), 1–66. doi: 10.1037/0033-295X.84.1.1

Schnotz, W., & Wagner, I. (2018). Construction and elaboration of mental models through strategic conjoint processing of text and pictures. *Journal of Educational Psychology*, 1–14. doi: 10.1037/edu0000246

Schreiber, D., Fonzo, G., Simmons, A. N., Dawes, C. T., Flagan, T., Fowler, J. H., & Paulus, M. P. (2013). Red brain, blue brain: Evaluative processes differ in democrats and republicans. *PLOS ONE, 8*(2), 1–6. doi: 10.1371/journal.pone.0052970

Schuldt, J. P., & Roh, S. (2014). Media frames and cognitive accessibility: What do 'global warming' and 'climate change' evoke in artisan minds? *Environmental Communication, 8*(4), 529–548. doi: 10.1080/17524032.2014.909510

Schulz, W. (2004). Reconstructing mediatization as an analytical concept. *European Journal of Communication, 19*(1), 87–101. doi: 10.1177/0267323104040696

Schwartz, S. A. (2014). The psychophysiology of politics. *Explore: The Journal of Science and Healing, 1*(1), 1–4. doi: 10.1016/j.explore.2014.08.014

Scott, J. C. (1998). *Seeing like a state: How certain schemes to improve the human condition have failed.* London, UK: Yale University Press.

Searle, J. (2010). *Making the social world: The structure of human civilization.* New York, NY: Oxford University Press.

Seife, C. (2014). *Virtual unreality*. New York, NY: Penguin.

Selby, J., Dahi, O. S., Fröhlich, C., & Hulme, M. (2017). Climate change and the Syrian civil war revisited. *Political Geography, 60*, 232–244. doi:10.1016/j.polgeo.2017.05.007

Sen, A. K. (1977). Rational fools: A critique of the behavioral foundations of economic theory. *Philosophy & Public Affairs, 6*(4), 317–344.

Sennett, R. (2012). *Together: The rituals, pleasures and politics of cooperation.* New York, NY: Yale University Press.

Sethi, S. P., & Schepers, D. H. (2014). United Nations global compact: The promise-performance gap. *Journal of Business Ethics, 122*(2), 193–208. doi: 10.1007/s10551-013-1629-y

Sevenans, J. (2017). How mass media attract political elites' attention. *European Journal of Political Research, 57*(1), 153–170. doi: 10.1111/1475-6765.12220

Shapiro, I. (2005). *The flight from reality in the human sciences*. Princeton, NJ: Princeton University Press.

Sherry, D. F., & Schacter, D. L. (1987). The evolution of multiple memory systems. *Psychological Review, 94*(4), 439.

Shortridge, J. E., & Guikema, S. D. (2016). Scenario discovery with multiple criteria: An evaluation of the robust decision-making framework for climate change adaptation. *Risk Analysis, 36*(12), 2298–2312. doi: 10.1111/risa.12582

Simon, H. A. (1956). Rational choice and the structure of the environment. *Psychological Review, 63*(2), 129–138. doi: 10.1037/h0042769

Singer, J., & McCarthy, T. (Writers). (2015). *Spotlight*. Beverly Hills, CA: Participant Media, First Look Media.

Sire, J. W. (2015). *Naming the elephant: Worldview as a concept*. Downers Grove, IL: Inter Varsity.

Smith, A. (1776). *An inquiry into the nature and causes of the wealth of nations*. New York, NY: Strahan and Cadell.

Smith, L. A., & Stern, N. (2011). Uncertainty in science and its role in climate policy. *Philosophical Transactions of the Royal Society, 369*(1956), 4818–4841. doi: 10.1098/rsta.2011.0149

Smith, N., & Leiserowitz, A. A. (2012). The rise of global warming skepticism: Exploring affective image associations in the United States over time. *Risk Analysis, 32*(6), 1021–1032. doi: 10.1111/j.1539-6924.2012.01801.x

Smith, N., & Leiserowitz, A. A. (2014). The role of emotion in global warming policy support and opposition. *Risk Analysis, 34*(5), 937–948. doi: 10.1111/risa.12140

Smith, R. (2014). Re-reading Plato: The slow cure for knowledge. In M. Papastephanou, T. Strand, & A. Pirrie (Eds.), *Philosophy as a lived experience: Navigating through dichotomies of thought and action* (Vol. 3, pp. 23–38). Zürich, Switzerland: LIT Verlag GmbH & Co. KG Wien.

Solms, M. (2014). *The neuropsychology of dreams: A clinico-anatomical study*. New York, NY: Psychology Press.

Speake, J., & LaFlaur, M. (Eds.). (1999). *The Oxford essential dictionary of foreign terms in English*. Oxford, UK: Oxford University Press.

Spearman, C. E. (1904). General intelligence, objectively determined and measured. *The American Journal of Psychology, 15*(2), 201–292. doi: 10.2307/1412107

Spearman, C. E. ([1927] 2005). *The abilities of man: Their nature and measurement*. Caldwell, NJ: Blackburn Press.

Spinoza, B. d. ([1883] 1955). *On the improvement of the understanding: The ethics; Correspondence* (R. H. M. Elwes, Trans.). New York, NY: Dover.

Staley, D. J. (2014). *Brain, mind and Internet: A deep history and future*. New York, NY: Palgrave Macmillan.

Stanovich, K. E. (1999). *Who is rational? Studies of individual differences in reasoning*. Mahwah, NJ: Psychology Press.

Stanovich, K. E. (2004). *The robot's rebellion: Finding meaning in the Age of Darwin*. Chicago, IL: University of Chicago Press.

Stanovich, K. E. (2009). *What intelligence tests miss: The psychology of rational thought*. New Haven, CN: Yale University Press.

Stanovich, K. E. (2012). On the distinctions between rationality and intelligence: Implications for understanding individual differences in reasoning. In K. L. Holyoak & R. G. Morrison (Eds.), *The Oxford handbook of thinking and reasoning* (pp. 343–365). Oxford: Oxford University Press.

Stanovich, K. E., & West, R. F. (2000). Individual differences in reasoning: Implications for the rationality debate? *The Behavioral and Brain Sciences, 23*(5), 645–665. doi: 10.1017/s0140525x00003435

Stanovich, K. E., & West, R. F. (2008). On the relative independence of thinking biases and cognitive ability. *Journal of Personality and Social Psychology, 94*(4), 672–695. doi: 10.1037/0022-3514.94.4.672

Stanovich, K. E., West, R. F., & Toplak, M. E. (2016). *The rationality quotient: Toward a test of rational thinking.* London, UK: MIT Press.

Steel, D. (2014). *Philosophy and the precautionary principle: Science, evidence, and environmental policy.* Cambridge, UK: Cambridge University Press.

Steffen, W. (2014). Managing Australia's environment in the Anthropocene. In D. Lindenmayer, S. Dovers, & S. Morton (Eds.), *Ten commitments revisited: Securing Australia's future environment* (pp. 227–235). Collingwood, Australia: CSIRO Publishing.

Steffen, W., Hunter, J., & Hughes, L. (2014). *Counting the costs: Climate change and coastal flooding* (pp. 1–83). Potts Point, Australia: Climate Council of Australia.

Stehr, N., & Ruser, A. (2017). Climate change governance and knowledge. In T. B. B. Scavenius & S. Rayner (Eds.), *Institutional capacity for climate change response: A new approach to climate politics* (pp. 28–42). New York, NY: Taylor & Francis.

Stehr, N., & von Storch, H. (2000). *Eduard Brückner: The sources and consequences of climate change and climate variability in historical times.* Dordrecht, the Netherlands: Springer Science.

Stern, N. (2007). *The economics of climate change: The Stern review.* Cambridge, UK: Cambridge University Press.

Stern, N., Bowen, A., & Whalley, J. (2014). *The global development of policy regimes to combat climate change* (Vol. 4). Singapore: World Scientific Publishing Company.

Sternberg, R. J. (2017). Theories of intelligence. In S. I. Pfeiffer (Ed.), *APA handbook of giftedness and talent* (pp. 145–161). Washington, DC: American Psychological Association.

Stevenson, A. (Ed.). (2014). *Oxford dictionary of English* (3rd edn). Oxford, UK: Oxford University Press.

Stevenson, K. T., Peterson, M. N., Bondell, H. D., Moore, S. E., & Carrier, S. J. (2014). Overcoming skepticism with education: Interacting influences of worldview and climate change knowledge on perceived climate change risk among adolescents. *Climatic Change, 126*(3), 293–304. doi: 10.1007/s10584-014-1228-7

Stiglitz, J. E. (2001). Foreword. In K. Polanyi (Ed.), *The great transformation: The political and economic origins of our time* (2nd edn). Boston, MA: Beacon Press.

Stiglitz, J. E. (2013). *The price of inequality.* London, UK: Penguin.

Stoknes, P. E. (2015). *What we think about when we try not to think about global warming.* White River Junction, VT: Chelsea Green Publishing.

Stollstorff, M. (2014). Genes of rationality: Building blocks for the neurobiology of reasoning. In W. De Neys & M. Osman (Eds.), *New approaches in reasoning research* (pp. 5–19). London, UK: Psychology Press.

Stollstorff, M., Bean, S. E., Anderson, L. M., Devaney, J. M., & Vaidya, C. J. (2013a). Rationality and emotionality: Serotonin transporter genotype influences reasoning bias. *Social Cognitive and Affective Neuroscience, 8*(4), 404–409. doi: 10.1093/scan/nss011

Stollstorff, M., Munakata, Y., Jensen, A. P. C., Guild, R. M., Smolker, H. R., Devaney, J. M., & Banich, M. T. (2013b). Individual differences in emotion-cognition interactions: Emotional valence interacts with serotonin transporter genotype to influence brain systems involved in emotional reactivity and cognitive control. *Frontiers in Human Neuroscience, 7*, 1–14. doi: 10.3389/fnhum.2013.00327

Stoner, A., & Melathopoulos, A. (2015). *Freedom in the Anthropocene: Twentieth-century helplessness in the face of climate change*. New York, NY: Palgrave Macmillan.

Strack, F., & Deutsch, R. (2004). Reflective and impulsive determinants of social behavior. *Personality and Social Psychology Review, 8*(3), 220–247. doi: 10.1207/s15327957pspr0803_1

Strömbäck, J., & Van Aelst, P. (2013). Why political parties adapt to the media: Exploring the fourth dimension of mediatization. *The International Communication Gazette, 75*(4), 341. doi: 10.1177/1748048513482266

Strong, M. F. (1992). *Report of the United Nations conference on environment and development: Rio de Janeiro*, 3–14 June, 1992. New York, NY: United Nations.

Suddaby, R. (2015). Can institutional theory be critical? *Journal of Management Inquiry, 24*(1), 93–95. doi: 10.1177/1056492614545304

Sun, R., & Zhang, X. (2004). Top-down versus bottom-up learning in cognitive skill acquisition. *Cognitive Systems Research, 5*(1), 63–89. doi: 10.1016/j.cogsys.2003.07.001

Sunstein, C. (2000). Deliberative trouble? Why groups go to extremes. *The Yale Law Journal, 110*(1), 71–120.

Schweppe, J., & Walters, M. A. (2016) *The globalisation of hate*. New York, NY: Oxford University Press.

Taleb, N. N. (2004). *Fooled by randomness: The hidden role of chance in life and in the markets*. New York, NY: Penguin.

Tamanini, J. (2016). *The global green economy index: GGEI 2016* (5th edn, pp. 1–58). New York, NY: Duel Citizen.

Tariq, A., Lempert, R., Riverson, J., Schwartz, M., & Berg, N. (2017). A climate stress test of Los Angeles' water quality plans. *Climatic Change, 144*(4), 635–639. doi: 10.1007/s10584-017-2062-5

Taylor, M. (2015). *The political ecology of climate change adaptation: Livelihoods, agrarian change and the conflicts of development*. London, UK: Routledge.

Taylor, P. M. (2003). *Munitions of the mind: A history of propaganda from the ancient world to the present era*. New York, NY: Manchester University Press.

Taylor, S. E. (1981). The interface of cognitive and social psychology. In J. H. Harvey (Ed.), *Cognition, social behavior, and the environment* (pp. 189–211). Hillsdale, NJ: Erlbaum.

Tesla, N. (1905). The transmission of electrical energy without wires as a means for furthering peace. *Electrical World and Engineer, 7*, 21–24.

Tetlock, P. E. (2005). *Expert political judgment: How good is it? How can we know?* Princeton, NJ: Princeton University Press.

Thatcher, M. (1988). Speech to the Royal Society. *Speeches, interviews and other statements.* Retrieved 25 June, 2014, from www.margaretthatcher.org/document/107346.

The Global Slavery Index. (2016). *The global slavery index*. Retrieved 2 August, 2017, from www.globalslaveryindex.org.

The Good Judgement Project. (2015). *The good judgement project*. Retrieved 11 May, 2015, from www.goodjudgmentproject.com/.

The Royal Society. (2017). *Climate updates: What we have learnt since the IPCC 5th Assessment Report*. Retrieved 5 February, 2018, from http://centaur.reading.ac.uk/74113/1/27–11–2017-Climate-change-updates-report-1.pdf.

Thuraisingham, M. (2013). *The secret life of decisions: How unconscious bias subverts your judgement*. Burlington, VT: Ashgate.

Tiffen, R. (2017). *Disposable leaders*. Sydney, Australia: NewSouth.

Tol, R. S. J. (2014). *Climate economics: Economic analysis of climate, climate change and climate policy*. Cheltenham, UK: Edward Elgar.

Toplak, M. E., West, R. F., & Stanovich, K. E. (2014). Assessing the development of rationality. In H. Markovits (Ed.), *The developmental psychology of reasoning and decision-making* (pp. 7–35). East Sussex, UK: Psychology Press.

Transparency International. (2016). *Corruption perceptions index 2015*. Retrieved 24 July, 2017, from www.transparency.org/news/feature/corruption_perceptions_index_2016.

Trenberth, K. (2012). Framing the way to relate climate extremes to climate change. *Climatic Change, 115*(2), 283–290. doi: 10.1007/s10584-012-0441-5

Tullock, G. (1974). *The social dilemma: The economics of war and revolution*. Blacksburg, VA: University Publications Blacksburg.

Tversky, A., & Kahneman, D. (1974). Judgment under uncertainty: Heuristics and biases. *Science, 185*(4157), 1124–1131. doi: 10.1126/science.185.4157.1124

Tversky, A., & Kahneman, D. (1981). The framing of decisions and the psychology of choice. *Science, 211*(4481), 453–458. doi: 10.2307/1685855

Tversky, A., & Kahneman, D. (1983). Extensional versus intuitive reasoning: The conjunction fallacy in probability judgment. *Psychological Review, 90*(4), 293–315. doi: 10.1037/0033-295x.90.4.293

Tyndall, J. (1861). The Bakerian Lecture: On the absorption and radiation of heat by gases and vapours, and on the physical connexion of radiation, absorption, and conduction. *Philosophical Transactions of the Royal Society of London, 151*, 1–36. doi: 10.1098/rstl.1861.0001

UNFCCC. (2015). *Synthesis report on the aggregate effect of the intended nationally determined contributions*. Bonn, Germany: UNFCCC. Retrieved 27 March, 2017, from http://unfccc.int/resource/docs/2015/cop21/eng/07.pdf.

United Nations Convention against Corruption. (2015). *United Nations convention against corruption: Signature and ratification status as of 1 December 2015*. Retrieved 23 May, 2016, from www.unodc.org/unodc/en/treaties/CAC/signatories.html.

Upin, C., & Hockenberry, J. (Writers) & C. Upin (Director). (2012). *Climate of doubt*. Arlington, VA: Public Broadcasting Service (PBS).

Vaish, A., Grossman, T., & Woodward, A. (2008). Not all emotions are created equal: The negativity bias in social-emotional development. *Psychological Bulletin, 134*(3), 383–403. doi: 10.1037/0033-2909.134.3.383

Valsiner, J. (2014). *An invitation to cultural psychology*. London, UK: Sage.

van der Pas, D., van der Brug, W., & Vliegenthart, R. (2017). Political parallelism in media and political agenda-setting. *Political Communication, 34*(4), 491–510. doi: 10.1080/10584609.2016.1271374

van Prooijen, J. W., & van Lange, P. A. M. (2014). *Power, politics, and paranoia: Why people are suspicious of their leaders*. Cambridge, UK: Cambridge University Press.

Varga, N. L., & Bauer, P. J. (2014). Conceptual knowledge extension: An examination of its development and the underlying cognitive processes involved. In R. Chen, *Cognitive development: Theories, stages and processes and challenges* (pp. 1–16). Hauppauge, NY: Nova Science.

Vaughn, L. (2015). *Beginning ethics: An introduction to moral philosophy*. New York, NY: W. W. Norton and Company.

Vinge, V. (1993). *The coming technological singularity: How to survive in the post-human era*. NASA. Lewis Research Center, Vision 21: Interdisciplinary Science and Engineering in the Era of Cyberspace, 11–22.

Visschers, V. H. M. (2018). Public perception of uncertainties within climate change science. *Risk Analysis, 38*(1), 43–55. doi.org/10.1111/risa.12818

Voegelin, E. ([1956] 2001). The collected works of Eric Voegelin. Order and history. In M. P. Hogan (Ed.), *The collected works of Eric Voegelin. Order and history*. Columbia, MO: University of Missouri Press.

von Mises, L. (1998). *Human action: A treatise on economics* (Scholar's ed.). Auburn, AL: The Ludwig von Mises Institute.

von Storch, H., & Stehr, N. (2006). Anthropogenic climate change: A reason for concern since the 18th century and earlier. *Geografiska Annaler: Series A, Physical Geography, 88*(2), 107–113. doi: 10.1111/j.0435-3676.2006.00288.x

Vygotsky, L. S., & Luria, A. R. (2013). *Studies on the history of behavior: Ape, primitive, and child* (J. E. Knox & V. I. Golod, Trans. J. E. Knox & V. I. Golod Eds.). Hoboken, NJ: Taylor & Francis.

Waddock, S., & Werhane, P. H. (2010). Finding wisdom within – the role of seeing and reflective practice in developing moral imagination, aesthetic sensibility, and systems understanding. *Journal of Business Ethics Education, 7*, 177. doi: 10.5840/jbee2010713

Walker, T. C. (2010). The perils of paradigm mentalities: Revisiting Kuhn, Lakatos, and Popper. *Perspectives on Politics, 8*(2), 433–451. doi: 10.1017/S1537592710001180

Wallace, A. R. ([1855–1859] 2013). *On the organic law of change* (J. T. Costa Ed.). London, UK: Harvard University Press.

Wardrop, A., & Withers, D. (Eds). (2014). *The para-academic handbook: A toolkit for making-learning-creating-acting*. London, UK: HammerOn Press.

Warhurst, J. (2014, August 12.). Entrenched corruption in NSW politics. *Sydney Morning Herald.* Retrieved 22 March, 2017, from www.smh.com.au/comment/entrenched-corruption-in-nsw-politics201408131032e4.html#ixzz3DSGDVBD5.

Wason, P. C. (1968). Reasoning about a rule. *The Quarterly Journal of Experimental Psychology, 20*(3), 273–281. doi: 10.1080/14640746808400161

Weaver, R. K. (2011). The politics of blame avoidance. *Journal of Public Policy, 31*(3), 371. doi: 10.1017/S0143814X00004219

Weber, M. (2009). *From Max Weber: Essays in sociology* (H. H. Gerth & C. W. Mills Eds.). Abingdon, UK: Routledge.

Webster, J., Malthouse, E., & Ksiazek, T. (2010). News-seekers and avoiders: Exploring patterns of total news consumption across media and the relationship to civic participation. *Journal of Broadcasting and Electronic Media, 54*(4), 551–568. doi: 10.1080/08838151.2010.519808

Weingast, B. R. (1997). The political foundations of democracy and the rule of law. *The American Political Science Review, 91*(2), 245–263. doi: 10.2307/2952354

Weiss, P. (1946). The quest for certainty. *The Philosophical Review, 55*(2), 132–151. doi: 10.2307/2181684

Weitzman, M. L. (2012). GHG targets as insurance against catastrophic climate damages. *Journal of Public Economic Theory, 14*(2), 221–244. doi: 10.1111/j.1467-9779.2011.01539.x

Welbourne, M. (2014). *Knowledge*. New York, NY: Routledge.

Wenger, E. C. (1998). *Communities of practice: Learning, meaning, and identity*. Cambridge, UK Cambridge University Press.

West, R. F., & Stanovich, K. E. (2015). A model for the assessment of rational thought and its potential operationalization. In R. Wegerif, L. Li, & J. C. Kaufman (Eds.), *The Routledge handbook of research on teaching thinking* (pp. 344–353). New York, NY: Routledge.

Westen, D. (2008). *Political brain: The role of emotion in deciding the fate of the nation*. New York, NY: Public Affairs.

Westland, G. (2015). *Verbal and non-verbal communication in psychotherapy*. New York, NY: W. W. Norton & Company.

White, M. J. (2012). *Political philosophy: An historical introductio*n (2nd edn). New York, NY: Oxford University Press.

White, R. (2011). *Transnational environmental crime: Toward an eco-global criminology*. New York, NY: Routledge.

Whitley, C. T. (2013). A picture is worth a thousand words: Applying image-based learning to course design. *Teaching Sociology, 41*(2), 188–198. doi: 10.1177/0092055X12472170

Whybrow, P. C. (2015). *The well-tuned brain: Neuroscience and the life well lived*. New York, NY: W. W. Norton.

Willer, D. (1967). *Scientific sociology: Theory and method*. Englewood Cliffs, NJ: Prentice-Hall.

Williams, L. M., Gatt, J. M., Schofield, P. R., Olivieri, G., Peduto, A., & Gordon, E. (2009). 'Negativity bias' in risk for depression and anxiety: Brain–body fear circuitry correlates, 5-HTT-LPR and early life stress. *NeuroImage, 47*(3), 804–814. doi: 10.1016/j.neuroimage.2009.05.009

Willis, P. (2013). *The ethnographic imagination*. Hoboken, NJ: Wiley.

Willmott, H. C. (2014). Science, governance and self-understanding: From anthropocentricism to ecocentrism? *Critical Policy Studies, 8*(1), 22–40. doi: 10.1080/19460171.2014.883858

Wilson, L., & Bauthus, K. (2014). *The art of eating through the zombie apocalypse: A cookbook and culinary survival guide*. New York, NY: BenBella Books.

Wilson, R. (2006). Policy analysis as policy advice. In M. Moran, M. Rein, & R. E. Goodin (Eds.), *The Oxford handbook of public policy* (pp. 152–168). New York, NY: Oxford University Press.

Wittgenstein, L. ([1922] 1999). *Tractatus logico-philosophicus* (C. K. Ogden, Trans.). New York, NY: Dover Publications.

Wolman, B. B. (1973). *Dictionary of behavioral science*. New York, NY: Van Nostrand Reinhold.

Wood, M. D., & Linkov, I. (2016). Science of mental modeling. In M. D. Wood, S. Thorne, D. Kovacs, G. Butte, & I. Linkov (Eds.), *Mental modeling approach: Risk management application case studies* (pp. 31–40). New York, NY: Springer.

World Health Organization. (2014). *10 facts about diabetes*. Geneva, Switzerland: WHO. Retrieved 12 July, 2017, from www.who.int/features/factfiles/diabetes/facts/en/.

Wright, E., Pegg, S., Ashfield, K., Davis, L., Frost, N., & Moran, D. (2004). *Shaun of the dead*. Universal City, CA: Universal.

Wright, R. (1995). *The moral animal: Evolutionary psychology and everyday life* (1st edn). New York, NY: Vintage.

Yaqin, A., & Morey, P. (2011). *Framing Muslims: Stereotyping and representation after 9/11*. New York, NY: Harvard University Press.

Young, O. R., (2017). *Governing complex systems: Social capital for the Anthropocene*. London, UK: MIT Press.

Young, S. T., & Dhanda, K. K. (2013). *Sustainability: Essentials for business*. Thousand Oaks, CA: Sage.

Zeineddine, F. B., & Pratto, F. (2014). Political distrust: The seed and fruit of popular empowerment. In J. Van Prooijen & P. A. M. Van Lange (Eds.), *Power, politics, and paranoia: Why people are suspicious of their leaders* (pp. 106–129). Cambridge, UK: Cambridge University Press.

Zevnik, L. (2014). *Critical perspectives in happiness research: The birth of modern happiness*. Dordrecht, the Netherlands: Springer.

Zhang, L. (2013). *The malleability of intellectual styles*. New York, NY: Cambridge University Press.

Zivin, J. G., Hsiang, S. M., & Neidell, M. J. (2018). Temperature and human capital in the short- and long- run. *Journal of the Association of Environmental and Resource Economists, 5*(1), 77–105. www.nber.org/papers/w21157.pdf.

Žižek, S. (2009). *First as tragedy, then as farce*. London, UK: Verso.

Zucker, L. G. (1977). The role of institutionalization in cultural persistence. *American Sociological Review, 42*(5), 726–743.

Zuckerman, M., Silberman, J., & Hall, J. A. (2013). The relation between intelligence and religiosity: A meta-analysis and some proposed explanations. *Personality and Social Psychology Review, 17*(4), 325–354. doi: 10.1177/108886831349 7266

Index

Page numbers in *italics* denote figures.

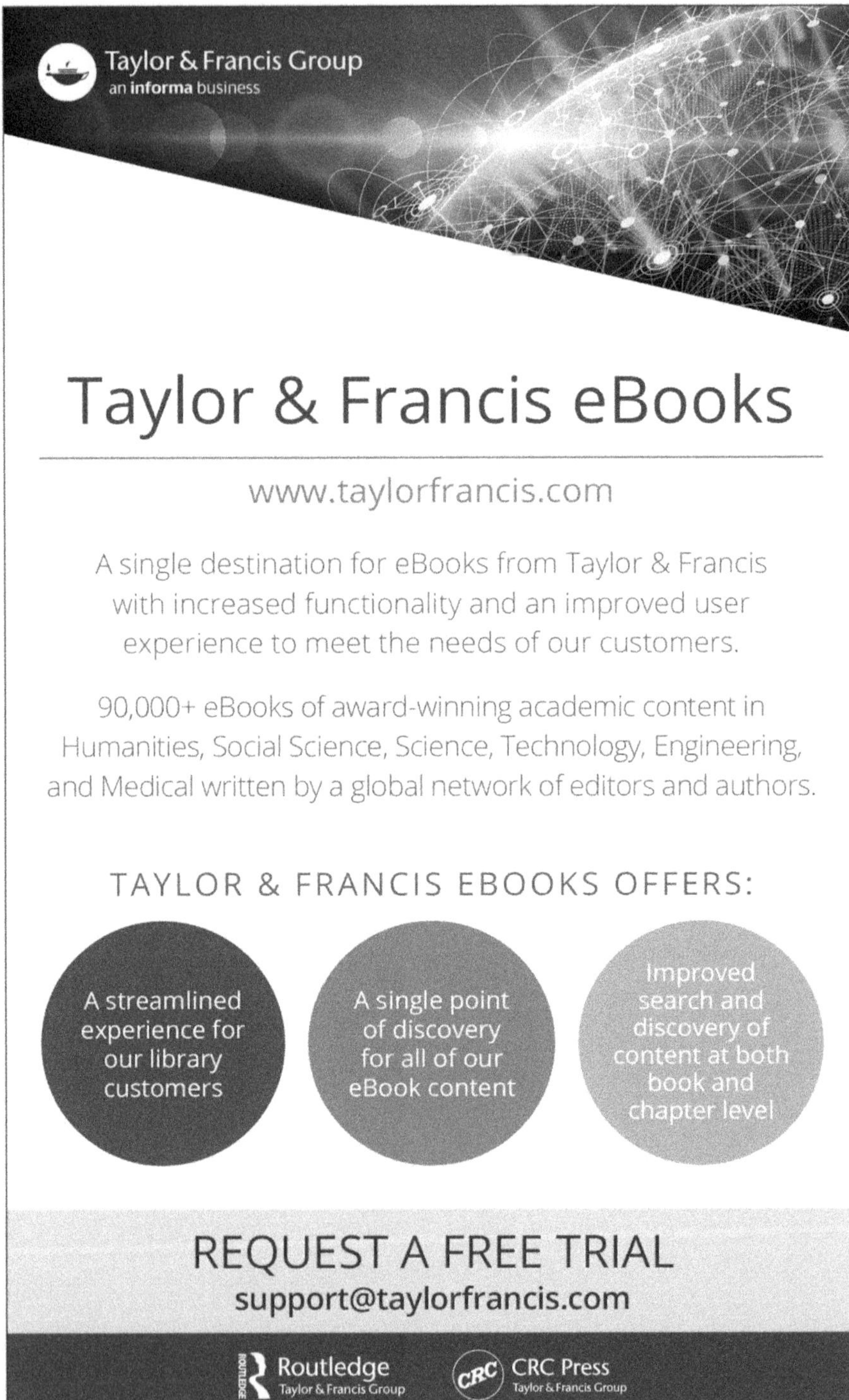
Taylor & Francis Group
an informa business
Taylor & Francis eBooks
www.taylorfrancis.com
A single destination for eBooks from Taylor & Francis with increased functionality and an improved user experience to meet the needs of our customers.
90,000+ eBooks of award-winning academic content in Humanities, Social Science, Science, Technology, Engineering, and Medical written by a global network of editors and authors.
TAYLOR & FRANCIS EBOOKS OFFERS:
A streamlined experience for our library customers
A single point of discovery for all of our eBook content
Improved search and discovery of content at both book and chapter level
REQUEST A FREE TRIAL
support@taylorfrancis.com
Routledge
Taylor & Francis Group
CRC Press
Taylor & Francis Group

For Product Safety Concerns and Information please contact our EU
representative GPSR@taylorandfrancis.com
Taylor & Francis Verlag GmbH, Kaufingerstraße 24, 80331 München, Germany

www.ingramcontent.com/pod-product-compliance
Ingram Content Group UK Ltd.
Pitfield, Milton Keynes, MK11 3LW, UK
UKHW021301190926
475701UK00009B/499

* 9 7 8 0 3 6 7 5 0 7 6 9 5 *